Right to Left:

The digital leader's guide to Lean and Agile

Mike Burrows

Foreword by John Buck

Published by New Generation Publishing in 2019

First Edition

Paperback: 978-1-78955-531-8
Ebook: 978-1-78955-532-5

www.newgeneration-publishing.com

New Generation Publishing

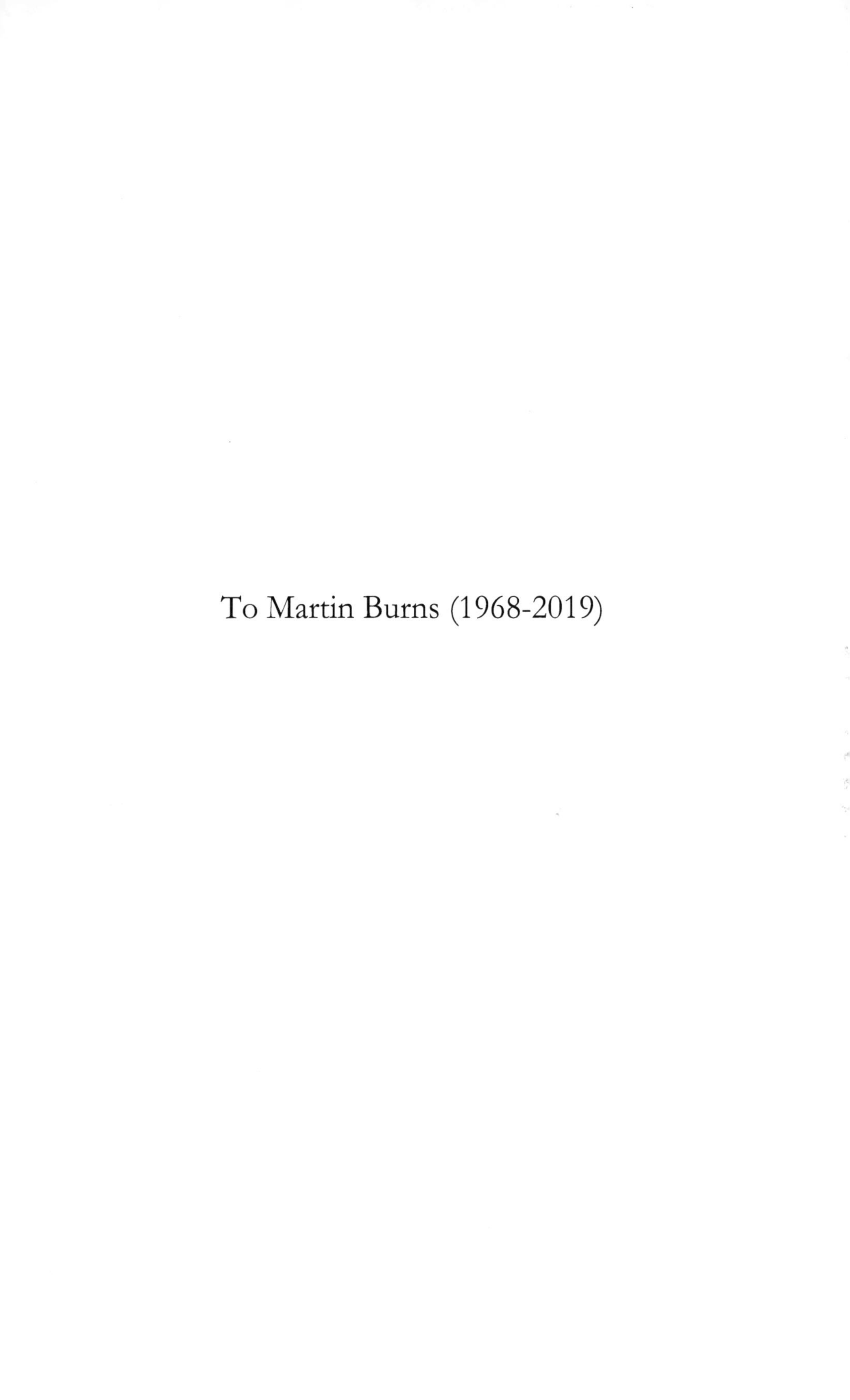

To Martin Burns (1968-2019)

Contents

Foreword

In *Right to Left*, Mike Burrows begins by pointing out that we are used to thinking of our companies' production processes as a causal sequence (typically drawn from left to right) that begins with inputs, moves through a series of transformations, and culminates in outputs in the form of products or services – typically operated as a *push process*. He then reverses that sequence and describes a *pull process,* in which needs and outcomes generate products and services and in turn create the inputs. For a long time now, this Lean approach has benefited bricks-and-mortar companies with much more efficiently-flowing supply chains.

This seemingly simple reversal in perspective with its focus on needs and outcomes has profound consequences for digital companies, too – benefiting their people as well their product development and service delivery processes. Collaboration between people of different specialties becomes easier, taking place in thriving cross-functional teams. Its purpose-driven perspective also makes it easier for diverse and supposedly competing management methods to "collaborate" (be integrated), which Mike demonstrates wonderfully through stories. Over several chapters he weaves together an impressive list of methods from the Lean-Agile landscape and beyond, including Scrum, Kanban, Lean, Agile, SAFe, XP, DevOps, Theory of Constraints, Wardley Mapping, Sociocracy, and Open Space.

The Right to Left reversal of perspective suggests other helpful reversals. One such is 'outside-in' strategic planning, in which strategies emerge from a desire to engage with the changing external environment rather than from a concern about what the organisation can do with its existing resources. Another is prompted by a question: *"How can we meet employees' needs for meaningful and fulfilling work?"* The typical result of this new perspective is bottom up (servant) leadership, 'intentful' communication, and focused collaboration aligned to the organisation's purposes.

Ultimately, Mike's book emerges from his focus on another outcome: meeting the needs of actual and aspiring leaders, needs that include reduced stress and deep collaboration and support from those they lead. As you read the book, be prepared to have some of your favourite mental frameworks delightfully upended!

John Buck
President, GovernanceAlive LLC
Silver Spring, MD, USA

Introduction

Did you know that there are two kinds of Agile? First, there's the kind capable of generating real passion in the shared experience of creating winning outcomes for customers. And then there's the kind that achieves only modest improvements in performance and delivers products that seem to delight no-one. Instead of the passion and engagement of the first kind of Agile, this second kind generates frustration at disappointing results and resentment that unfamiliar ways of working have been implemented for so little benefit.

A disparity this wide suggests that something fundamental must be happening. You don't need to have been around the Agile community for long to hear people complaining about it; perhaps you've noticed it for yourself first hand. Many put it down to Agile "going mainstream" and becoming the victim of its own success, as though there's something inevitable and perhaps unstoppable about its decline. I find this explanation unhelpful however – not because there's no truth in it, or because it's pessimistic (I'm an optimist by nature), but because the insight doesn't lead in any useful way to action.

Let me offer a complementary and more helpful explanation. That first kind of Agile – the passion-generating kind – is characterised by a shared focus on outcomes. The second kind – the kind that dances with disengagement – is characterised by a focus on implementation, with designs, working practices, and workload largely determined up front, often with little meaningful involvement from the majority of people who will be tasked with carrying it out. Hardly Agile at all, by most meaningful definitions.

Exploring the differences between these two styles led me to *Right to Left,* the central metaphor of this book. *Right to Left* is shorthand for consistently, deliberately, and even provocatively starting with outcomes – with needs being met – and working backwards from there, keeping outcomes always in the foreground. It's a visual metaphor, and to understand it, just imagine a diagram of the delivery process drawn conventionally (to Western eyes at least) with inputs on the left and outputs on the right. Now put outcomes even further to the right; it's from there that our thinking starts, and it's to there that we're constantly drawn.

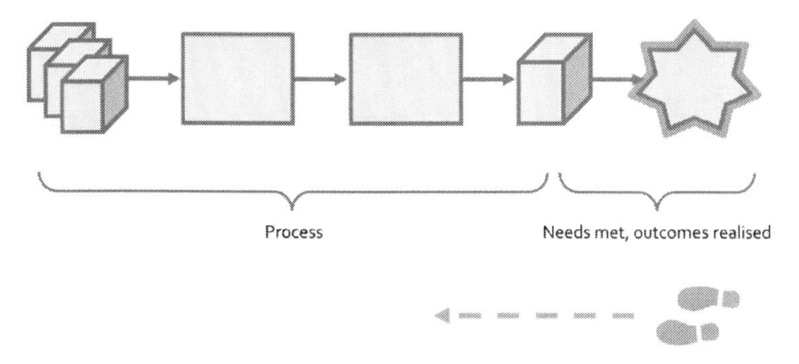

Figure 1. Approaching a process right to left

Right to Left

To see how the metaphor works, let's try it with something familiar. Imagine you had the job of explaining all you could about Lego®. Where would you start? Would you start:

1. "From the left", with truckloads of plastic feedstock arriving at the Lego factory, or
2. "From the right", with children playing with the finished product

Most people would find it natural to choose a starting point somewhere towards the right, at (or at least near) the *outcome*, not the *input*. I certainly would!

If you are able, now do your best to describe how Agile works. Again, where would you start? That's harder, but try these:

- From the left, with backlogs, planning activities, and so on
- From the right, with people collaborating over the rapid evolution of working software that is already beginning to meet needs

If you have the knowledge and experience to recognise both of these descriptions, now ask yourself these two rather awkward questions:

1. Which of them do you think will best help people understand Agile as a fundamental departure from 20th century-style project delivery?
2. Which version is the one most often told?

Oops! To the practitioner community I say this: every time Agile starts from the left, it seriously undermines its own message, and I fear for the long-term consequences. Sometimes it does more immediate harm; more on this in chapter 4, which includes a comparison of left-to-right and right-to-left approaches to Agile adoption.

So what does *Right to Left* mean in practical terms? For now, think of it as the discipline to consider outcomes (and there are several different kinds of those) before solutions, ends before means, vision before detail, "why" before "what", "what" before "how", and so on. It can also mean considering outputs before inputs, but give me outcomes over outputs, every time.

As we dig into this more deeply, a couple of frequently-occurring concepts will seem to appear at both ends, left and right:

1. *Need*: On the right, the most meaningful outcomes are those about which you can say with confidence that someone's need was met[1]. But before needs can be met, they must be identified, prioritised, and their respective outcomes described, activities that take place towards the left.

2. *Assumption*: On the right, confidence increases every time an assumption is validated; invalidating a well-framed assumption generates learning, so that has real value too, even when the result is not the one that was hoped for. As with needs, assumptions must first be surfaced, articulated, and prioritised. The sooner this can be done, the sooner they can be tested, perhaps without going to the trouble and expense of building a complete solution.

The shorter the gap in time and the closer the collaboration between the activities at the two ends, the more effective those activities are. At the extreme, it's a continuous process, one in which we're meeting needs and learning through delivering, all at the same time, and all the time.

The digital leader's guide to Lean and Agile

Do you see in digital technology the opportunity to serve your customers better, to meet their needs more effectively? Do you recognise (or at least suspect) that this may have profound implications for how your organisation should work? Do you want to help make that happen?

Whether or not you consider yourself a technologist, if your answer to those questions is "yes", you are what we refer to in this book as a digital leader. If you are a digital leader, aspire to be one, or think that sometime soon you might need to become one, then this book is for you.

Or perhaps you're here primarily to feed your interest in Lean and Agile.

Whatever your current level of knowledge, this book is for you too, especially if you're interested also in organisation design and leadership. You will find here both an accessible guide to the Lean-Agile landscape and through the Right to Left metaphor a helpfully challenging perspective on it. Our digital scope might not coincide exactly with yours, but it's rich with authentic examples not only of Lean-Agile practice but of right-to-left thinking too.

Digital leadership, Lean, Agile, and outcome-orientation come together in a delivery approach that makes the most of the unprecedented flexibility, immediacy, and ubiquity of web-based, mobile, and similar technologies. These digital technologies make it both remarkably easy and increasingly necessary to discover and meet the ever-changing needs of users on a continuous and ongoing basis. That's a game-changer, not just in the sense of a dramatic change in fortunes, but in the sense that we're choosing to play a completely different game, a game so different that it shouldn't be explained in the terms of the old one.

For a generation or more, managers were taught that software development is a linear process, one that should be planned out in advance, and whose success depends mainly on how diligently those plans are executed. But consider these challenges:

- The response of the market to any new digital product or service is hard to gauge until it's in the hands of real users
- User engagement is hard to control; competitive environments change, perceptions evolve, and in this age of hyperconnectivity, people talk to each other!
- Competition is fought on the basis not only of product features, but on a range of broader attributes such as the quality of the overall service experience, speed of product evolution, and future potential

Any one of these very real challenges would be sufficient to seriously undermine that old, linear model. Accepting that, we have come to embrace the idea that there is something inherently complex[2] about our work. Invoking complexity in this way is not a boast: we're not claiming for example that we're solving technical problems any more challenging than those solved by traditional IT projects. Rather, it's a humble acknowledgement that we're not in complete control of the future, and that to pretend otherwise is to invite failure.

This might sound like bad news, but it needn't be taken that way. Granted, it's a serious threat to organisations still clinging to the belief that the way to create a digital product is to spend long periods of time and large amounts of money before they come out of hiding and launch the finished article on an unsuspecting market. But it's fantastic news to organisations willing to recognise that digital products and services present a wonderful opportunity to interact with and learn about the outside world. Embracing this, they set themselves up to stimulate and incorporate this learning on a continuous basis – starting not with the finished product but with tests of product viability, adding enhancements and making refinements over time, guided by real-world observation.

This iterative, emergent, experiment-based approach isn't just for garage-based startups or tech giants such as Google, Amazon, and Netflix. For example, I've had the privilege of witnessing from the inside some of the dramatic changes happening within the UK government sector[3]; seen even from the outside as a citizen user of digitally-offered government services, it's impressive. With genuine respect for their achievements, I say that if organisations as set in their ways as governments can do it, any organisation can. Your organisation can!

But to do anything new takes leadership. Digital leadership can take many forms, but let's get down to specifics. Two key and interdependent roles typically form the primary axis of product or service evolution:

- Product leadership – continuously exploring the problem space, discovering and prioritising needs that the product or service must somehow meet, carrying the product vision
- Technical leadership – ensuring that at every stage of its evolution, the product or service gets the kind of the technical realisation it deserves, carrying the technical vision

This is a creative collaboration, and the sign that it's working effectively is that it's more than the sum of its parts, not only producing things that couldn't be produced working alone, but achieving things that might not even have been conceived otherwise. In that sense, it's a great model for other collaborative working relationships. This extends even to the customer relationship, a relationship so vital that it is specifically called out in the founding document of the Agile movement, the Agile Manifesto[4].

Other leadership roles in areas such as design, delivery management, process, and organisational change become more important as teams grow in size. It's not unusual for these to be filled by external experts (or 'practitioners'); if you're one of those, this book is for you too, and I hope you find something new and interesting in it.

A business of any real size will of course have other kinds of leader, including people in traditional leadership roles in functions such as sales, marketing, HR, and finance. This book has less to say about those, but if you're in such a role and your organisation is about to embark on something exciting in the digital space, I hope that you too will find it valuable.

Digital, Lean, Agile, and Lean-Agile

Tom Loosemore, the author of the UK government's first digital strategy and a former deputy director of the UK's Government Digital Service (GDS), defines digital as follows:

> **Digital**: Applying the culture, processes, business models & technologies of the internet era to respond to people's raised expectations[5]

Wherever my usage of the word allows this definition, you may safely assume that I intend it. Implicit is the deeper insight that digital is about more than technology; it is about being part of a dynamic social system that is changing itself before our eyes.

Lean and Agile have a great deal to contribute in this space. We'll cover their most relevant concepts in chapters 1 and 2 respectively, but to place them in some historical context:

- Lean owes its inspiration to a motor manufacturing company – Toyota – and its management system, the Toyota Production System (TPS). It is Lean that gave us terms such as *just-in-time*, *stop the line*, and *kanban*, all of which are about managing and improving *flow*[6], delivering maximum value to the customer with the minimum of delay, interruption, or rework.
- Agile burst into life through the publication of the *Manifesto for Agile Software Development*, popularly known today as the Agile Manifesto. This was the work of a group of practitioners that met in 2001 at the mountain resort of Snowbird, Utah, USA; its signatories were there as representatives of what were then known as 'lightweight' software development methods. The manifesto took the form of a set of values and principles that both unified these emerging methods and explained what set them apart from the conventional wisdom.

Lean and Agile in combination – Lean-Agile – isn't quite so well defined, but I do find a home in its community. For my second book, *Agendashift*, I described Lean-Agile with the phrase *"Celebrating Lean and Agile, both separately and together"*. This worked well enough then and it will do for the moment; we will shoot for a more formal definition in chapter 2.

In chapters 3 and 4 we'll introduce several of the branded frameworks that identify with or are complementary to Lean, Agile and Lean-Agile. There are enough of these that I couldn't mention them all; if I omit a favourite of yours, please accept that no slight is intended. Some of the best-known frameworks invite both right-to-left and left-to-right interpretations, a clue that the way they are understood and adopted matters greatly, perhaps more than the choice of framework! We'll return to that potentially subversive thought in chapter 4.

Chapter overview

This book is organised into six chapters, the first four of which have a strong right-to-left theme:

1. **Right to left in the material world** – introducing Lean, the strategic pursuit of flow
2. **Right to left in the digital space** – introducing Agile and Lean-Agile
3. **Patterns and frameworks** – popular Lean, Agile, and Lean-Agile frameworks and how they combine and complement each other
4. **Viable scaling** – the Agile scaling frameworks, organisational viability, and the challenges of change

The last two chapters approach questions of organisational design and leadership from angles complementary to that core theme:

5. **Outside in** – strategy and governance in the 'wholehearted' organisation
6. **Upside down** – Servant Leadership and its relationships to collaboration and adaptability in the supportive, 'intentful', customer-focussed organisation

From chapter 2 onwards we will pay some quick visits to Springboard DIY, a fictitious retail company (unrelated to the real store I visit in chapter 1) that serves as a rough composite of some of the organisations I have led, advised, or served in other ways in the past decade or so. The actual organisations involved cover a very wide range, from the voluntary and government sectors through to energy and finance; each has its special nuances but many of their challenges recur with such regularity that any initial surprise soon wears off! However, rather than dwell on their obstacles, I find it much more productive to focus on what opportunities lie beyond them, and it is those that Springboard largely represents.

From chapter 3 onwards most noticeably, some technical terms are capitalised. This is done not for reasons of emphasis but simply to respect the conventions of the communities from which these terms or their particular definitions are drawn.

At the end of each chapter you'll find some questions for reflection, both as a reminder of what you have just read, and to help you identify some of your organisation's opportunities for change. If you'd prefer just to skim them and move on to the next chapter that's completely fine – they are brought together at the end of chapter 6 and Appendix A for a more structured exercise. Either way, less important than their answers are the thoughts that they may provoke.

Appendix B, *"My kind of…"*, is not technical glossary, but it gathers together some informal definitions that are especially characteristic of this book, all of them applicable in the context of digital leadership and most of them helpful in wider contexts too.

Related

You need not have read either of my previous books in order to enjoy *Right to Left* – in fact they're written for different audiences, even if they overlap somewhat. My first book, *Kanban from the Inside,* was published in 2014 and was the first values-based treatment of Kanban – meaning both the Kanban Method and the tool from which the method took its name. My 2018 book, *Agendashift: Outcome-oriented change and continuous transformation*, describes a right-to-left approach to change and transformation; I had *Right to Left* already in mind as I wrote it, and many readers will find this most recent book the most accessible starting point of the three. With the benefit of hindsight, my personal journey from values via outcomes to *Right to Left* does make sense, but I had no idea of what was in store when I made my first step with a career-changing blog post written over the 2013 New Year holiday[7]!

Before we move on to the first chapter, some pointers that I don't want to leave until the end of the book:

- This book has an online home at agendashift.com/**right-to-left**. If you have feedback of any kind you can easily reach me there. You'll also find links to resources, recommended reading (a selection of the references provided in each chapter's endnotes), and the blog.
- Readers will find a warm welcome in the #right-to-left channel in the Agendashift Slack (agendashift.com/**slack**)
- If you're liking what you're reading, an appreciative tweet to @asplake (me), @Right2LeftGuide (this book), &/or hashtag #Right2LeftGuide would be wonderful, thank you.

Now, enjoy!

---

[1] This is my deliberately provocative *Definition of Done*: *"Someone's need was met"*

(agendashift.com/**done**).

[2] When the words *complex* or *complexity* appear in this book, they are used as understood by the various schools of *complexity theory*, of which the best-known in the Lean-Agile community is *Cynefin* (en.wikipedia.org/wiki/Cynefin_framework). Briefly, complex systems aren't just *complicated* – things you can disassemble and reassemble – but have behaviours that emerge and evolve over time through the relationships and interactions between things. Feedback loops, delays, memories, and so on make complex systems very hard to predict. These effects are very apparent when human nature is involved, for example in social systems such as organisations and economies.

[3] The British newspaper *The Guardian* described the UK's Government Digital Service (GDS) as *"revolutionising the way that 62 million citizens interact with more than 700 services"* and *"pioneering the way that large complex corporations reinvent themselves"*. See *Government Digital Service: the best startup in Europe we can't invest in*, Saul Klein, www.theguardian.com/technology/2013/nov/15/government-digital-service-best-startup-europe-invest (theguardian.com, 2013)

[4] You can read the Agile Manifesto – or to use its full name, the *Manifesto for Agile Software Development* – at agilemanifesto.org. The particular phrase I have in mind here is *"Customer collaboration over contract negotiation"*.

[5] *Definition of Digital*, definitionofdigital.com

[6] I highlight *flow* because it's such an important concept. Be assured however that it is a metaphor so powerful than you can and should trust your intuitive understanding of the word. To take an example from nature, along the length of a great river you might see different kinds of flow: smooth flow (impressive in its own quiet way), white water rapids (lots of energy on show but much of it lost), eddies (in which some of the movement is backwards), and backwaters (hardly any movement at all); for all of these effects their counterparts in business are easy to find. They might go by different names there, but our experience of them can be much the same.

[7] *Introducing Kanban through its values*, blog.agendashift.com/2013/01/03/introducing-kanban-through-its-values/

Chapter 1. Right to left in the material world

Disaster! You were just moments away from completing that DIY project, just one more hole needed, but your electric drill finally broke down. Your work has come to a standstill; further progress will be impossible until you find a replacement.

Deep breath… Not actually a disaster. Checking on your phone, you find that your nearest DIY store has just the kind of drill you need. You place an order for immediate collection, jump in your car, and it's not long before you're enjoying your completed project. Whew!

What just happened?

At one level, your little story seems rather mundane, a simple transaction. You had a need, and the purchase of a product from a store helped to meet that need. If, however, you're at all curious about how your order was fulfilled and care to dig deeper, there is so much more to tell:

- Your store has capabilities that didn't exist just a few years ago, enabling you to collect without delay an item that has already been picked from the shelf or warehouse on your behalf
- Via its global distribution system, your store sources the latest products from a range of suppliers, maximising your chances of finding something that meets your needs at a price that makes sense to all parties involved
- Without interrupting the production of its existing product range, the manufacturer has over recent weeks enhanced its production systems so that this new model can be assembled efficiently
- Hand in hand with the manufacturer, suppliers have responded to new demands, changing not just quantities, but specifications, packaging, and delivery protocols

- Your product incorporates the latest advances in battery, motor, and control technology, those the fruit of countless experiments conducted at the frontier of knowledge in engineering, materials science, electronics, physics, and chemistry

In short, you've managed to buy what turns out to be a remarkable piece of brand new technology at a price acceptable to you, thanks to a very modern kind of supply chain.

Decades ago, conventional wisdom would hold that affordability was only achievable through mass production, economies of scale bringing costs down to levels that consumers can afford. But in our new era of rapid innovation and consumer choice, efficiencies of scale can't explain what just happened in our story. Consumers don't want yesterday's products. Retailers don't want to hold stock that will soon be obsolete and hard to sell. Manufacturers don't want to produce stock in quantities that retailers might not shift quickly.

If what we're seeing isn't mass production, something else must be going on, something that delivers a different kind of efficiency. Let's try to understand what makes it all possible.

We begin by visualising the supply process just described, drawing a simple *value stream map*. A value stream map is a visual representation of the *value stream*; the value stream describing the *flow* (i.e. the movement) of work products of various kinds between and through what we hope are *value-adding activities* (meaning that they transform those work products in ways that help to make them more valuable to you, the eventual customer).

Understanding the value stream, right to left

What do we know? First, there's the *output* of the value stream – the drill you needed – which you collected from the DIY store. Unless they think to ask you why you made your purchase, hidden from the store are the *outcomes*: your last hole drilled – an urgent need met – and your project completed, the point of it all (Figure 2):

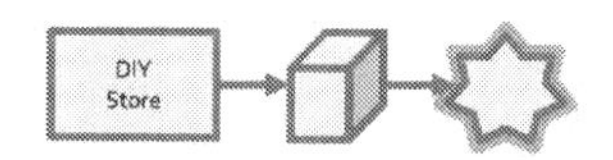

Figure 2. The end of the process, its output, and its outcomes

What else do we know (or can at least guess)? We can reasonably assume that some kind of distribution system brought the drill to your store (Figure 3):

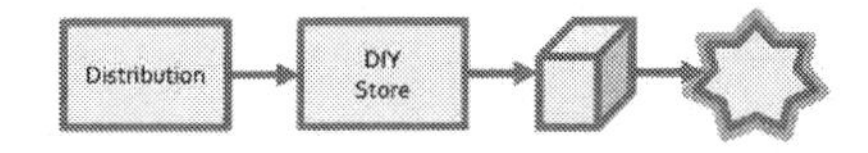

Figure 3. Distribution

Your drill was assembled by a manufacturer (Figure 4):

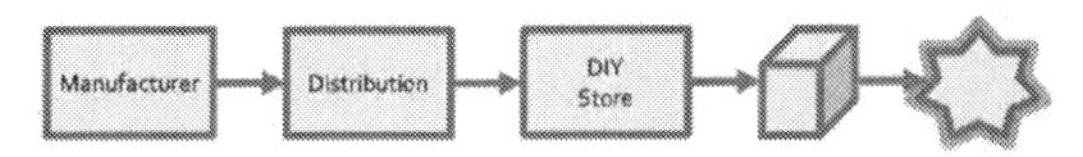

Figure 4. The manufacturer

The manufacturer sourced parts and raw materials from its suppliers[8] (Figure 5):

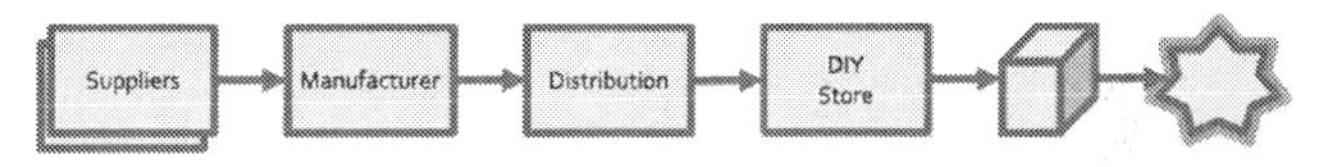

Figure 5. Suppliers

Innovations were developed by the manufacturer, its suppliers, and in other research establishments, universities, for example (Figure 6):

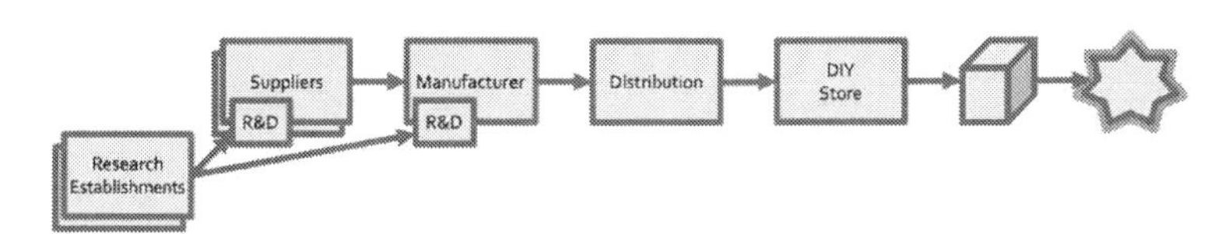

Figure 6. Research and development

If we wished, we could go further back, showing the suppliers' suppliers, their suppliers, and so on, potentially not just a straightforward flow but an entire *value network*[9]. We will stop here, however.

Especially in the western world, we are so used to seeing value stream maps and other process maps drawn with arrows pointing from left to right that it's easy to miss that they're often constructed right to left. If they weren't explicitly taught to do so, practitioners of *value stream mapping* soon discover for themselves that this is almost always the most effective tactic. It's not hard to see why: anchoring the process on a key moment of value creation and following it upstream keeps the map focused on things that matter. At each step in the process, only these three things need to be established:

1. The activity performed here
2. The resources needed for that activity to be performed successfully

3. Where those resources come from, the next places to look

Were you to start instead from the left – with raw materials and other basic supplies – you would find it much more difficult to draw your map. You would discover any number of activities that are performed on those things and all the places they can go to immediately thereafter, and very quickly the range of possibilities explodes. Even assuming that you chose the best place to start, your search for the path that leads to the outcome of interest will be an inefficient one.

Go and see

Our first value stream map was very high level, and not (I will freely admit) based on exhaustive research. For our next one, we're going to do some real discovery work on the ground. To use a popular Lean idiom whose intention is to warn against armchair management, we shall *go and see*[10].

Specifically, let's see how the 'click & collect' process works at the DIY store, quite literally on the shop floor of a store in my home town. With Cara, the store's manager, I start at the collection desk and work my way upstream, understanding the process from right to left:

Mike: So I arrive here at the collection counter to pick up the drill I ordered. Then what happens?

Cara: You identify yourself to the member of staff behind the counter and they hand you your drill along with the receipt that has already been printed. Done!

Mike: That's easy! What happens before then to make that all possible? Where did everything come from?

Cara: When you place your order, it flashes up on the screen here.

(As she points to the screen, a new order – not mine – happens to pop up in bright pink)

Cara: Whoever is on duty here clicks on it, prints out the customer's receipt, our paperwork, and a label that helps us identify your items quickly when you arrive. They then go and pick up your items from wherever they're held – the shelves, a secure cupboard (for a high-value item like your drill), or perhaps the warehouse.

(We walk to the power tools aisle)

Mike: I see tools on display, some in boxes, and some locked away as you described. How can you be sure that you'll have the item I ordered?

Cara: The IT system keeps track of how much should be in stock, but the real work is done in the 'gap walk'. At regular intervals we walk the whole store, and scan the bar codes on these shelf labels whenever we spot either that stock is below the minimum amount for display or if we're overstocked for some reason. Stock then gets replenished from the warehouse as needed. We do the gap walk there too, and once we're ready to order an appropriate quantity, the order goes out on my approval. Display quantities and order quantities may vary seasonally, but the process remains the same.

Notice something interesting: nothing in this process moves without a signal from downstream, 'downstream' meaning closer to me, the eventual customer:

- My order notifies the central IT system of my intention to pick up the drill soon
- The IT system notifies staff at the click & collect counter in the store of my choosing that my order has arrived; they then print the paperwork and do the stock picking
- The need for replenishment is determined visually by staff at the shelves and in the warehouse (not virtually, by an accounting system)
- Orders to suppliers go out when available stock is seen to have fallen to its trigger level, a parameter that along with order quantities may vary over the course of the year

A simple but effective system that is both robust (self-correcting) and flexible (tuned to the dynamics of each individual product line, of which there are thousands).

The factory floor is a very different setting but the story is similar. Subassemblies get made just before they're needed to go into the final product, not batched up in advance. Limited quantities of more basic supplies are kept at each workstation in small bins, replenished as they near depletion.

The signal that a workstation is in need of some kind of input is often given in the form of a card, known as a *kanban*[11] in Japanese and now quite widely in the West too. In the typical factory floor *kanban system* (we'll see another kind in chapter 3), the kanban are sent upstream from a workstation facing a potential shortfall; the required inputs flow downstream in response, arriving when needed. Multiple workstations can be connected in this manner, all the way across the factory.

Across the whole value stream, instead of large buffers of stock, long production runs, and warehouses kept full of finished goods, we see activities happen in response to specific demand. Instead of *resource efficiency*[12] – products built speculatively in large batches by people and machines kept maximally busy in a bid to keep unit costs low – we're seeing *flow efficiency*[13]. Flow efficiency minimises the time that elapses between inputs being committed into the system and its outputs being consumed. Equivalently, it maximises the proportion of time that work products spend undergoing value adding activities.

This focus on that new kind of efficiency represents such a radical departure from traditional ways of thinking that it needs a name. We call it Lean.

What kind of Lean?

Depending on who you talk to, Lean seems to mean several things:

1. A body of knowledge inspired by the Toyota Production System (TPS)
2. A set of tools, many of them with Japanese names or obscure acronyms
3. A cost-cutting approach, "doing more for less"
4. A framework for continuous improvement

Unfortunately, while there is truth in each of the above, none of them is an entirely satisfactory basis for a definition of Lean. In summary, each has serious problems:

- Definitions 1 & 2, focussed on Toyota's systems and tools, make the mistake of confusing the Lean of today for its origin story. Undoubtedly, Toyota's practices at the time were better than those that were then the norm, but they and world move keep moving on, and our job is not to entrench practices that are now decades old.
- Definition 3, focussed on cutting costs, confuses Lean with just one of its potential outcomes. Cost control has its place of course, but to focus primarily on cost tends to take the organisation in a very un-Lean-like direction, one that prioritises resource efficiency at the expense of flow efficiency.
- Definition 4 – Lean as synonymous with continuous improvement – isn't so much wrong as woefully inadequate, leaving out too much of what is necessary to be successful.

Yes, Lean is indeed inspired by the Toyota Production System. And rightly so: the Toyota story is remarkable, well worth studying, and the company itself has been generous in facilitating that study. The TPS is a vital part of that story; so too are its tools. But blind copying doesn't work; Toyota themselves describe this behaviour as embarrassing! Most of us don't work for motor manufacturers, and even if we did, by the time we had finished our implementation, our smarter competitors would have moved ahead of us long ago. Were you to try it, not only would you be replicating the tools of the past, but you'd be doing it ignorant of whether they will be effective in your particular context. Get that wrong, and the results will be horrible.

Does "more for less" adequately describe the rise of Toyota from the ruins of post-war Japan to its position as one of the leading motor manufacturers in the world? Of course it doesn't. And it's worse than inadequate: it's highly misleading. When your focus is on squeezing the maximum out of existing resources, flow tends to get worse, not better.

And yes, Lean involves continuous improvement[14]. It's essential! Continuous improvement inside Toyota is highly developed, formalising how they make changes ranging from small production line tweaks to large restructurings. The problem is that continuous improvement runs into two serious problems when it is implemented on its own:

1. We've all seen continuous improvement initiatives run out of steam, and the truth is that continuous improvement is not naturally self-sustaining. Yes, the textbooks describe improvement as happening in cycles[15], but it is not the case that one improvement must lead inevitably to another.
2. Unfortunately for the lazy manager, continuous improvement is no substitute for strategy. It's not enough just to solve problems; they have to be the right problems. For the right problems to be identified and addressed relentlessly, there must be a shared understanding of the outcomes to be achieved and awareness of the obstacles that stand in the way of those outcomes.

If this book's kind of Lean is none of those things exactly, then what is it? In order both to understand Lean both in its historical context and to apply it successfully today, these two essential insights must be recognised:

- **Insight 1**: It is in the mutual interest of both customers and suppliers to improve flow. Not only does faster and smoother flow improve the customer experience, it makes a wide range of efficiencies possible and creates opportunities that would not otherwise exist.
- **Insight 2**: The pursuit of flow is a serious, long-term endeavour. No

single change can be guaranteed to deliver perfect results – in fact the majority of experiments deliver more in learning (revealing new obstacles to overcome) than they do in performance. This has some important implications for organisational design. At every level of the organisation there will be a need for:

- **Engagement** – involving people with first-hand knowledge of where both the obstacles and the opportunities lie
- **Leadership** – guiding and continuing the journey even when each step is surrounded by uncertainty
- **Development** – equipping staff and leaders, ensuring that the next generation will continue a journey whose exact destination can't be known today

Toyota was the first to internalise these insights. In highly distilled form they're encapsulated in the twin 'pillars' of *Just in time* (JIT) and *Respect for people*, and they underpin TPS. Unfortunately, what first caught everyone's attention wasn't the insights, but just some of the tools Toyota had developed in order to act on them, the tools most visible to outside observers at that time.

In *This is Lean*[16] – written far enough into the 21st century not to repeat the mistakes of the past – authors Niklas Modig and Pär Åhlström describe Lean simply as *"a strategy of flow efficiency"*, certainly a big improvement on the kind of descriptions offered at the beginning of this section. Inspired by their description but seeking to make it less dependent on a technical term, I offer this definition:

Lean:

> The strategic pursuit of flow, a deliberate process of organisational learning

Rather less succinctly, here's a version that tries to convey the degree of organisational challenge involved:

Lean:

> The pursuit of flow as strategic imperative, an open-ended and purpose-driven endeavour that continuously and deliberately engages people at every level of the organisation in a learning process

Lean passes all the tests of good strategy[17]: it identifies real challenges, it provides a clear framework for decision making, and the actions it generates tend to reinforce each other. Demonstrably, it is also a strategy for the long term: Toyota has been committed to theirs for decades already and they're not done yet.

Several kinds of leadership are required to implement this strategy successfully:

- **Product leadership**, conceiving and evolving products that meet the needs of willing customers and are rewarding to use
- **Technical leadership**, designing products that can be built, delivered, and supported effectively
- **Market leadership**, connecting people and products, managing demand
- **Process leadership**, finding better ways to operate and manage the delivery process
- **Change leadership**, catalysing change through sponsorship, experimentation, facilitation, coaching, and coordination
- **Executive leadership**, removing structural impediments, ensuring that the organisation pulls together, now and into the future

It is rare to find all of these dimensions fully developed and in perfect balance in one person. Lean is therefore very much a team sport, even at the leadership level.

To finish this chapter, two important components of the Lean body of knowledge: a description of the Lean improvement process in the form of the Lean principles, and a model of the kinds of obstacles (or 'wastes') that the Lean improvement process typically must tackle.

Lean principles

The five principles of Lean were identified as long ago as 1991 in *The Machine That Changed the World*, the book by Womack, Jones, & Roos that brought the term *lean production*[18] to prominence. As they're introduced, notice just how "right-to-left" they are:

Principle 1: Identify value

We begin by identifying value from the customer's perspective. In the story at the opening of this chapter, value was created when:

1. You collected your drill from the store, completing a transaction
2. You were able then to meet your underlying need, your need to drill that last hole and complete your project

You won't identify value by looking only at the production process. To fully appreciate it, you need to look beyond the delivery or the transaction and discover what makes it valuable in the hands of the customer. You will enter the territory of needs and outcomes, and it's helpful to learn its language.

Principle 2: Map the value stream

Working backwards from those key moments of value creation, map out the steps of the process, distinguishing activities that add value from the customer's perspective from those that don't. If non-value-adding activities – *waste*[19] – can be eliminated or minimised, flow will improve.

Principle 3: Create flow

Encourage work to flow smoothly through the process towards the customer, so that each work item progresses without delay through each activity and between activities. If you're not doing so already, now is also a good time to start measuring system performance, establishing a baseline against which future performance can be compared.

Principle 4: Establish pull

For a while, you might create flow through a combination of careful vigilance and a readiness to intervene whenever work stops flowing. You'll gain some idea of what the process is capable of, but as a long-term strategy it is neither reliable, cost-effective, nor sustainable. In fact, all that interference may actually make things worse[20]!

A much more elegant, efficient, and robust approach is to establish *pull*. Instead of 'pushing' work into the system without regard to system capacity and then dealing with the consequences for flow, the system is designed to 'pull' work into the system when it is ready. *Pull systems* allow work to enter an activity only when two conditions are met:

1. A demand signal has been received from the direction of the customer, so that nothing is made prematurely or unnecessarily
2. The activity itself has capacity available, ensuring that this new work won't inflict further delay on a part of the system that is already suffering

When multiple pull systems are connected together, we see these demand signals cascade rapidly in a right-to-left direction. In order to complete their work, work cells close to the customer – downstream, or on the right – request whatever items they need from cells upstream, to their left. Those cells in turn request what they need from further upstream, and so it continues up the line, right to left. The work itself flows left to right, passing smoothly through activities that are operating continuously within safe and effective limits of load.

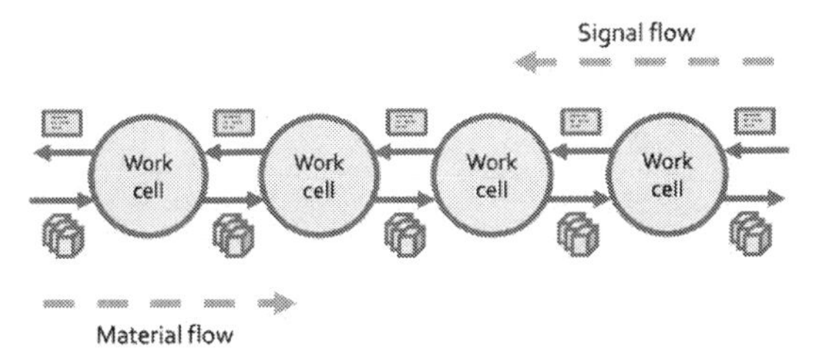

Figure 7. A kanban system spanning multiple work cells

Kanban systems of the kind described in this chapter represent an important class of pull system. A kanban system's design, policies, and parameters determine how many signal cards – the kanban – can be in circulation at any given time, a deliberately-introduced constraint that establishes pull and improves flow.

Principle 5: Pursue perfection

As pull becomes established, flow begins to happen more naturally, and *impediments to flow* become much easier to see. To the trained eye, waste is everywhere; Lean practitioners often joke that their inability to "unsee" waste is a curse!

An opportunity for a significant change of management focus is created: instead of helping work through the system and dealing with specific issues as they arise, managers are now tackling systemic issues, addressing underlying flaws in the design of systems. The scope of this improvement activity can be surprisingly broad and challenging: the 'system' isn't just the physical production line, but the way it is operated, managed, supplied, staffed, developed, and so on. This work must be done in the face of external pressures, social dynamics, and other factors that are impossible completely to control, making it truly a job for the long term. Consultants can get you started or help you over a hump, but ultimately the pursuit of perfection must be owned internally and made a recognised part of every manager's duties.

These principles can be summarised as *"Understand, manage, and improve flow, right to left"*:

- **Understand** the delivery process, starting from the right, where needs are met
- **Manage** flow, embracing the mantra *"Stop starting, start finishing"*,

helping to internalise flow and pull, and building trust in the process

- **Improve** the process by addressing systemic impediments to flow, tackling aspects of system design or behaviour that get in the way of work being completed and meeting a need

In the next chapter we'll see how this structure applies to digital, not just as deliberate tactics, but ingrained habits.

The 7 wastes

Lean's *7 wastes* help to classify the various impediments to flow typically encountered when pursuing flow in the material world:

- **Waste 1: Transportation** – moving material from A to B (perhaps via C) when to do so adds no customer value.
- **Waste 2: Inventory** – keeping more work in the system than necessary, the result of starting work prematurely or in oversized batches. Excess inventory causes delay, must be financed until value is realised, and introduces a host of second-order effects such as poor quality and the inability to understand the system's true potential.
- **Waste 3: Motion** – inefficiencies introduced by poor ergonomics – tools and materials disorganised or out of reach, uncomfortable and even unsafe working environments.
- **Waste 4: Waiting** – work delayed behind other work (see Inventory above) or by the failure of supplies or other essential resources to arrive in time.
- **Waste 5: Overproduction** – producing something before it is needed, an act that may seem harmless but is frequently a root cause of excess inventory and its related problems.
- **Waste 6: Over-processing** – performing activities that are unnecessarily elaborate, time-consuming or expensive, using equipment more sophisticated than necessary.
- **Waste 7: Defects** – the double, triple, or quadruple whammy! Effort is spent first on producing defective work, then on detecting the problem, and finally on rectifying it. It gets worse still – sometimes orders of magnitude worse – when the first person to notice the problem is your customer.

It's important to recognise that these are the wastes of Lean *production*. Away from the production of material goods – in product development, service delivery, or digital for example – some of these wastes can seem far less relevant. Even as metaphors they can mislead: in the virtual world for example, what possible meanings could we attach to **transportation** and **motion**, and are they wasteful in every case?[21]

Some wastes are universal however:

- Excess **inventory** is simultaneously a cause and a symptom of waste
- Whether you're **waiting** for something physical or virtual, the impact is much the same
- By definition, **over-production**, **over-processing**, and similar excesses must be wasteful
- **Defects** are so insidious that it usually pays to *"stop the line"*[22] and deal with them at source the moment they are discovered

Inventory is so fundamental that it features in a key mathematical relationship, *Little's law*[23]. Averaged over suitable periods of time (and with surprisingly few other mathematical caveats):

Time in process = Inventory / Throughput

Where:

- *Time in process* is the time taken for similar units of work to pass through the system (or part thereof)
- *Inventory* is the number of those units of work in the system
- *Throughput* is the rate at which those units pass through the system per unit of time

Once you grasp Little's law, Lean's apparent obsession with inventory seems only sensible! If you can reduce inventory without unduly impacting throughput, time in process must also reduce. If you can eliminate delays, you will see inventory reduce or throughput increase (and perhaps both).

Not counted among the 7 wastes but in important categories of their own[24]:

- *Variation* – inconsistency, unpredictability, unevenness, or irregularity – whether of demand, flow, quality or some other measure of system or product performance
- *Overburdening* – people and systems over-stressed and unable to perform at their best, the result of excess inventory, unexpected demand, management failure, or some other cause

As championed by the likes of Lean Six Sigma[25] and the great W. Edwards Deming[26], the process improvement community generally understands variation to be an important problem to tackle in its own right. Certainly it's true that understanding and addressing systemic sources of variation can be a powerful way to improve a process.

In two very real senses, the desire to eliminate variation seems entirely compatible with Lean:

1. The more smoothly a process runs, the less inventory it needs to hold as insurance against disruption
2. Predictability can be of value to the customer in its own right

On the other hand:

1. It would be foolish to eliminate variation whenever it's a source of significant business opportunity, customers willing to pay a significant premium for their needs to be met in different (and perhaps innovative) ways
2. Predictability can be an expensive substitute for speed; "slow and steady" doesn't always win the race!

The key is to recognise and exploit the valuable forms of variation at the same time as striving to eliminate the rest. Toyota provides a powerful example in its seemingly perverse choice to build a range of different products on the same production line in what seemed to be impossibly small batches. To deal with the unpredictability and delays this tactic introduced, they forced themselves to innovate new techniques such as *rapid changeover* and *single-minute exchange of die*[27]. Faced with the apparent trade-off between predictability and opportunity, they chose a path of flexibility, achieving the best of both goals.

The 8th waste

The name *"7 wastes"* seems to have stuck, but there is in fact a widely-recognised 8th waste, variously worded as **untapped creativity** and **unfulfilled human potential**. This is a late addition, and it must be said that in the eyes of some Lean practitioners its dissimilarity to the first 7 wastes causes it to sit rather awkwardly in that list.

Whether or not it belongs with the others, it does reveal some of the intent behind *Respect for people* and speaks to another of Toyota's deep insights. The alignment of interests in the pursuit of flow is based on more than just economics:

- Success depends on people up and down the value stream showing curiosity, creativity, and an appetite for experimentation

- People can find meaning and enjoyment in their work when appropriately challenged, empowered, and equipped
- Customers and even society can benefit in undreamt-of ways when the right products continue to be conceived, created and delivered at the right time.

There are some serious challenges to leaders here. Too often, initiative is suppressed, too much reliance is placed on financial and other extrinsic rewards to the detriment of more intrinsic sources of motivation, and the opportunity to innovate is lost through lack of vision or a desire to control. These are reminders – if any were needed – that most waste should be blamed not on the people doing the work, but on the design of the working environment. If blame should be assigned to people at all, start with the leaders responsible. Better still, focus on the systems that train those managers to be that way!

Closely related to the 8th waste is the concept of *psychological safety*[28]. People won't give their best if they fear that this will lead to negative consequences. Teams will never reach their full potential if team members expect new ideas to be received judgementally. Few will attempt anything challenging if failure is widely regarded as something to fear. Psychological safety is very much a team phenomenon, but it's one in which leaders set a strong example. If your teams don't have it and you want to see improvement, start there.

Reflections

1. How do you understand those *"key moments of value creation"* that take place "on the right", happening in and resulting from the interactions between your organisation and your customers?
2. Working from right to left, how do you understand your business's value streams – the processes that culminate in those key moments?
3. More generally, how do managers in your organisation maintain an up-to-date understanding of their value streams and appropriate awareness of what is happening in them on the ground? To what extent is this based on first-hand observation?
4. Working again from right to left, how do activities in your value streams coordinate with their counterparts upstream so that their needs are met in good time?
5. What would *"a strategy of flow efficiency"* look like for your organisation? What would sustain it?
6. Where and how do Lean's 7 *wastes* – transportation, inventory,

motion, waiting, overproduction, over-processing, and defects – impact your value streams today?

7. By what mechanisms, policies, or levers are inventory, throughput, and/or time in process controlled or influenced?

8. How is psychological safety cultivated in your teams?

---

[8] You will notice that some of our boxes describe organisational units rather than activities. Sloppy! When you're working at such a high level of detail it's sometimes best to choose the most obvious names for things, names that people will identify with. As you drill down, it's important to identify activities accurately so that the map shows what actually gets done. Organisational boundaries can always be drawn on later.

[9] Clayton Christensen coins the term *value network* in his 1997 book on disruptive innovation, *The Innovator's Dilemma: When New Technologies Cause Great Firms to Fail*, Clayton M. Christensen (Harvard Business Review Press, reprint edition 2013).

[10] In Lean circles, you may also hear the Japanese for "go and see", *genchi genbutsu*. Semi-Anglicised terms you may also encounter are *go to the gemba*, meaning "go to the actual place" – the place in question typically being the shop floor – and *gemba walk*, which is Lean shorthand for going to the actual place and walking around, looking for improvement opportunities. See en.wikipedia.org/wiki/Gemba.

[11] Whether referring to a signal card (in the singular) or signal cards (plural), the Japanese word *kanban* is the same.

[12] Technically, *resource efficiency* refers to the proportion of time a resource is kept busy, known also as its *utilisation*. Typically this is expressed as a percentage, a utilisation of 100% describing a resource that is always busy, never idle.

[13] Technically, *flow efficiency* measures the proportion (percentage) of time that the product spends actively being worked on in ways that add value, relative to its overall elapsed time through the system. A flow efficiency of 100% would imply perfectly smooth flow, with value being added to the product the whole time it moves through the process.

[14] The Japanese term most often used for continuous improvement is *kaizen*, literally "change for better".

[15] See en.wikipedia.org/wiki/PDCA for the granddaddy of all improvement cycles. Call me a heretic, but I made the decision long ago to stop teaching improvement cycles, preferring instead to teach the framing of improvements as experiments and the organisation of ongoing improvement-related work as two distinct concepts. Describing something as a cycle does not automatically mean that it will sustain

itself!

[16] *This is Lean: Resolving the Efficiency Paradox*, Niklas Modig & Pär Åhlström (Rheologica Publishing, 2014)

[17] *Good Strategy/Bad Strategy: The difference and why it matters*, Richard Rumelt (Profile Books, 2011)

[18] Credit for first coining the terms *lean* and *lean production* goes to John Krafcik, whose article *Triumph of the lean production system* was published in the Sloan Management Review in 1988.

[19] *Muda* in Japanese, literally "futility" or "uselessness"

[20] Beware of what W. Edwards Deming called *tampering*, intervening with insufficient knowledge of the impact. See curiouscat.com/management/deming/tampering.

[21] To give credit where it's due, *Lean Software Development: An Agile Toolkit*, by Mary Poppendieck & Tom Poppendieck (Addison-Wesley Professional, 2003) did dwell on the 7 wastes and was highly influential and a milestone in in my own Lean-Agile journey.

[22] *"Stop the line"* is the reference to Toyota's practice of empowering workers to call a process to a halt if they suspect any kind of problem. On the production line, this is done through the physical act of pulling the *andon cord*.

[23] en.wikipedia.org/wiki/Little%27s_law

[24] In Japanese *variation* and *overburdening* are *mura* and *muri*. "Unevenness" is the more literal translation of *mura* and it does sometimes get used in the Lean literature; here we follow Deming and use "variation". *Muda*, *mura*, and *muri* often appear together, referred to collectively as the 3M model (not to be confused with the company of that name).

[25] en.wikipedia.org/wiki/Lean_Six_Sigma

[26] en.wikipedia.org/wiki/W._Edwards_Deming, and I highly recommend his book *The New Economics* (MIT Press, 2nd Edition, 2000)

[27] en.wikipedia.org/wiki/Single-minute_exchange_of_die.

[28] See en.wikipedia.org/wiki/Psychological_safety and the landmark Google study *What makes a Google team effective*, rework.withgoogle.com/blog/five-keys-to-a-successful-google-team/. Spoiler: psychological safety was identified as the #1 factor.

Chapter 2. Right to left in the digital space

We're at the offices of Springboard DIY, an online retailer of home improvement products. Alex, a senior developer, has just found product manager Rowan in the corridor. Alex sees that Rowan is about to leave for an external meeting.

Alex: I'm so glad I caught you before you left. We were about to make those power tools changes permanent but there are some things I'd like to check with you first. Can we spend a few minutes on it together sometime soon?

Rowan: Sure! is there is anything I should be concerned about? Everything I've seen in the data since that release looks great so far – more people are navigating faster through those pages and they're making purchases!

Alex: Don't worry, that's what I'm seeing too. I can't help wondering though if there's an opportunity here that we're missing. It might make sense to look into it while it's still fresh in our minds.

Rowan: Sounds intriguing! I must run now, but how does 2 o'clock this afternoon sound? I'll be back in the office by then.

What just happened?

We just saw a little example of the kind of thing you can expect to see when people are collaborating closely over the rapid evolution of working software – working in the important sense that it's already beginning to meet needs. It's what we should expect to see emerging very quickly in any organisation that places high value on these things:

- **Collaboration**, both internal (within and between teams) and external (with customers most especially)
- **Continuous delivery** – working software (or working products and services) that already meets some kind of need, continuing to improve through rapid successions of small *incremental* (additive) enhancements and *iterative* (repeated) refinement[29], and the infrastructure to achieve this rate of change efficiently[30]

- **Adaptability** – not just flexibility over priorities and schedules, but allowing the designs of product, process, and organisation to change in response to (or in anticipation of) changing circumstances and new knowledge, not least the knowledge stimulated by the delivery process

In a development environment, three processes are happening simultaneously:

1. The relatively predictable work of delivering well-understood changes through a mainly linear process, outcomes achieved quickly in single passes through the system
2. Developing the product through an emergent and knowledge-building process that has the potential to influence everything it touches in unpredictable ways; outcomes achieved patiently through disciplined experimentation
3. Developing the infrastructure of technology, process, and culture needed to support the first two

A key difference from the typical factory environment is that it's the same people involved in all three processes. In a healthy product development environment, these three processes are kept in balance. When they are allowed to get out of balance, the consequences can be serious:

1. When the focus is on implementing predetermined requirements and insufficient attention is paid to how well each piece of finished work actually meets needs, the almost inevitable result is mediocrity.
2. Learning at the expense of delivery leads to products that don't sustain themselves. When at the expense of infrastructure, the result is superficiality and fragility – products that are perhaps beautiful on the surface but built on shaky foundations.
3. Over-emphasising infrastructure leads to "all framework, no end product" – over-engineered systems, interaction with the real world delayed to the point of irrelevance. When infrastructure is under-emphasised, it usually means that problems are being stored up for the future, the accumulation of *technical debt* that will one day need to be repaid.

How well this works in practice of course depends greatly on the people involved, their respective abilities and responsibilities, and their respect for each other. At Springboard DIY, being fully aware of the consequences of getting this balance wrong, they work hard to get it right. In her product role for example, Alex is a strong champion for her products, their users, and their respective needs. At the same time, she fully understands her inter-reliance on colleagues in business, delivery, technology, and design roles. Everyone in the leadership team acknowledges that they can't achieve much of any significance on their own, and they know through experience that the real innovations tend to come in the collaborations. Alex has seen it enough times that she trusts it to happen and has learned to enjoy the process, however unclear the way forward may seem at the beginning. Ask any of her peers, and their experience would be similar.

What kind of Lean-Agile?

In chapter 1, we saw some of the quite different ways in which Lean is understood. Before we get to Lean-Agile, let me describe this book's kind of Agile, a kind of Agile that should already sound familiar:

Agile:

> People collaborating over the rapid evolution of working software that is already beginning to meet needs

Expanding a little:

Agile:

> People bringing their various skills to bear on the rapid evolution of working software that is already beginning to meet needs, working in teams that place high value on collaboration and adaptation

That's my highly condensed and "from-the-right" interpretation of the Agile Manifesto (agilemanifesto.org), worded to describe a sweet spot for digital. If you want to know what Agile is, the manifesto is where you need to start. Agile isn't a defined process, method, or framework; Agile means embracing manifesto values – values of collaboration, working software, and adaptability.

The Agile movement exists because the manifesto values have resonated with many people. They see how they might play out in meaningful conversations, the opportunity to build things that actually work for people, and the ability to keep improving the working environment to the mutual benefit of developers, customers, and the organisation. In other words, they speak to some basic human needs that not every employer satisfies.

For a values system to be more than just wishful thinking however, there must be a clear relationship between the values, the kinds of behaviours expected, and the assumptions that underpin these behaviours. In the case of Agile, these assumptions are well documented – not least by the manifesto itself – and they go a long way to describe the behaviours:

- **Assumption 1**: Direct, ongoing collaboration with customers is necessary to develop and maintain a mutual understanding of needs and potential solutions; in complex environments both can be expected to evolve over time as they interact

- **Assumption 2**: Collaboration between people working across the entire process is what makes the whole greater than the sum of its parts – not just multiple perspectives brought to bear on complex problems, but new ideas created and refined in the interactions

- **Assumption 3**: The most effective way to build anything complex is to start with something that works, and ensure that it stays working as it evolves [31]. This is true not just for products, but for the working environment in terms of its technical infrastructure, processes, practices, organisation, culture, and external relationships.

Agile's breakthrough is the result of bringing these assumptions together to everyone's attention in the form of a compelling values statement. The underlying message is clear: wherever those assumptions are likely to hold it would be smart to behave accordingly, even if that demands a radical departure from previous ways of working.

The idea of customer needs being met through small increments and frequent iteration does seem to suggest something flow-like. Indeed, a popular piece of Agile jargon is *"flow of value"*. Does this then imply that Agile is nothing other than *"the strategic pursuit of flow"*, no different to the Lean of chapter 1? That turns out to be a controversial question; people I respect have answered it both positively and negatively:

- Yes – how could it be anything else?

- No – Agile places some things above flow (to good or bad effect, depending on your point of view)

Redefining Agile in Lean terms would not only fail to resolve this question, many would regard it as an impertinence. Instead, we have Lean-Agile, a way to honour a joint heritage. It might be defined as follows:

Lean-Agile:

The strategic pursuit of flow in complex environments, the organisation placing high value on collaboration, continuous delivery, adaptation, and learning

It's worth comparing that definition with my two previous definitions for Lean and Agile, the former given in chapter 1 and the latter at the beginning of this section:

Lean:

The pursuit of flow as a strategic imperative, an open-ended and purpose-driven endeavour that continuously engages people at every level of the organisation in a learning process

Agile:

People bringing their various skills to bear on the rapid evolution of working software that is already beginning to meet needs, working in teams that place high value on collaboration and adaptation

Some Agilists would have no difficulty embracing all three of these definitions and reconciling them with their understanding of Agile. Others would take the quite justifiable position that Agile is mainly about teams or mainly about software and baulk at that prospect. Stepping back from that team level perspective however, more questions are raised:

- Given that organisations both inhabit and contain complex environments – competitive landscapes, social systems, etc – shouldn't they all seek to be Agile in some sense?
- Is it possible to make Agile work not just at the level of development teams but at the scale of the organisation? In other words, is it possible to *scale* Agile?

In a sense, this book is about taking those questions seriously. In order to do that, we're going to start with a working system that is already beginning to meet needs – Springboard DIY's digital division – and explore the kinds of models, methods, tools, and patterns that help both to put it in place and to sustain its continued evolution. You can be the judge of what might have been gained or lost along the way.

Go and see

So let's return to Springboard DIY and see what fast and smooth flow looks like. While we're there, we'll sketch out its value stream map.

Naturally, we'll start from the right. But where is that exactly? To take the perspective of one of Spingboard's customers, we might try to define it as the moment an online transaction is completed. Unfortunately, this choice has two problems:

- Customers often visit Springboard DIY online for reasons other than to make purchases
- Springboard's staff aren't involved on a per-transaction basis, making the business process uninteresting for our current purposes,

even if it might be interesting for reasons such as customer experience and technical architecture

There is however another important kind of value that digital services are especially good at capturing. There is immense business value in understanding what customers want and how they prefer to interact. When delivery or development processes end with an explicit step of *validation* designed to inform the process that precedes it, we call the process *validated learning* (a term from Lean Startup, which we will cover in the next chapter). The output of validation is a stream of *insights*; when insights are internalised and acted upon, one very important outcome is *learning*.

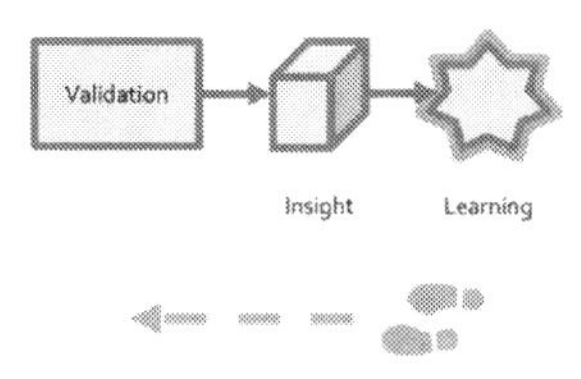

Figure 8. Validation

Springboard's commitment to validated learning explains the conversation between Alex and Rowan at the top of this chapter. They have been running an experiment, field-testing some changes to the power tools section of their site. Using a technique called *A/B testing*, these changes weren't deployed in one go, but released to a small subset of the user population and then rolled out progressively as confidence grew. For experiments like these, certain metrics are identified in advance as indicators to the success or otherwise of the change. Operating at the scale of a big retailer, statistical significance can be reached in a matter of hours or days, and the change made permanent if it is judged to be beneficial, or rolled back otherwise.

Because the team at Springboard has been working like this for a long time, many measurements get taken, second by second, user interaction by user interaction. That's a lot of data, plenty for everyone! Product specialists look at product trends and financial metrics; user experience (UX) specialists look at how (and how quickly) users move through the site; technical specialists look at system performance metrics, ensuring that the system is behaving as expected and has the capacity to cope with peaks in demand. Effective validation depends on knowledge of current baseline levels of the relevant metrics and some explicit assumptions about how they might be impacted. Or to put it another way, good experiment design looks ahead to validation.

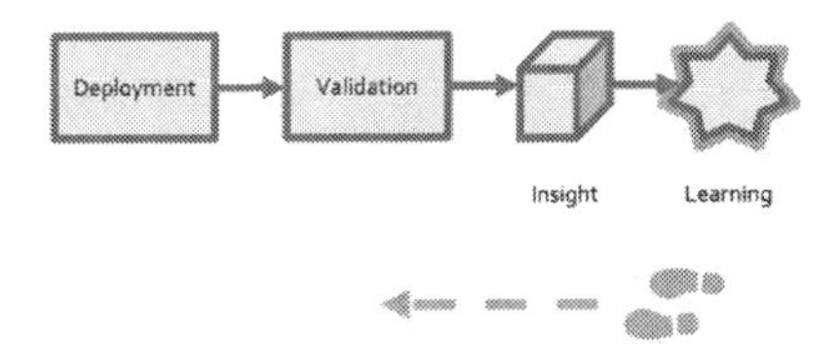

Figure 9. Deployment

To be effective, validated learning needs fast feedback. For fast feedback to be practically and economically viable, it must be fast, easy, and cheap to get changes deployed, the practical side of continuous delivery. Happily, fast and easy also means low risk, because the amount of change contained in any release is small enough to be known to everyone involved, and because the deployment process itself is highly repeatable. At Springboard, deployments are regarded not as special events but business as usual; away from critical periods of peak customer demand (bank holiday weekends for example) they're happy to make multiple deployments per day if the opportunity is there.

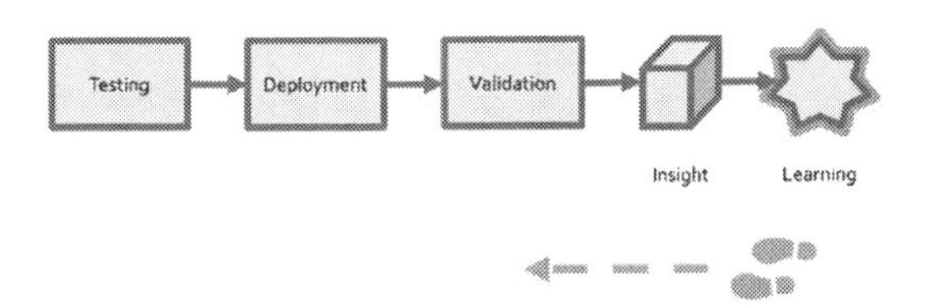

Figure 10. Testing

It should go without saying that Springboard's digital team is not in the habit of deploying changes that haven't undergone a number of checks. That said, their QA specialists don't spend the majority of their time in the traditional activities of manually testing new functionality or coordinating signoffs; by the time most changes reach them, functional testing and integration testing has not only been completed, it has been automated so that any future regressions will be caught immediately. Instead, they're mostly collaborating with developers on changes still in their early stages, engaged in exploratory testing, measuring system performance, testing on behalf of users with accessibility needs, and so on – making sure that everything about the online experience has kind of consistency and polish that customers have come to expect. When they don't have customers with them, they can at least try to walk in their shoes.

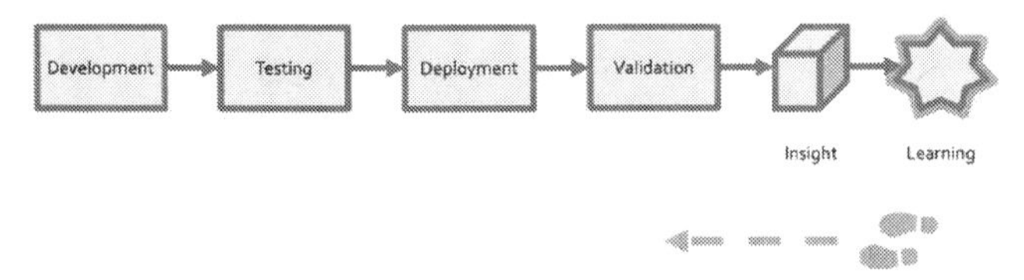

Figure 11. Development

Development at Springboard DIY is a multi-disciplinary endeavour, involving deep specialisms (server-side infrastructure, user interfaces, and everything in between), craftsmanship, and a great deal of collaboration. No team member works entirely independently – in fact many spend the majority of their time with others, perhaps *pairing* for extended periods with another team-member, *swarming* in larger groups over particular challenges, or collaborating with colleagues from other teams. "Working software" is the constant focus; changes are integrated throughout the course of the day, developers receiving feedback on their work both as they make them (via tests they write in parallel with their changes and execute repeatedly as they work) and as their changes are shared with their colleagues (via tests run on shared servers, integrating with other services).

All of this work is done looking forward to the moment – typically no more than days away – when it is in the hands of customers, doing something useful, "working" in the truest sense. Part of the technical challenge – and also the opportunity of digital – comes in instrumenting it so that the team will get rapid feedback on how well each change performs from a user perspective.

Good data brings further opportunities, for example the ability to make personalised recommendations, which is to the benefit of users if done well. However, data must be managed with great care. Privacy and information security are designed not only into the Springboard's online offering at a technical level, but also into the policies that govern how it is designed, built, and operated. *Respect for people* extends beyond the organisation!

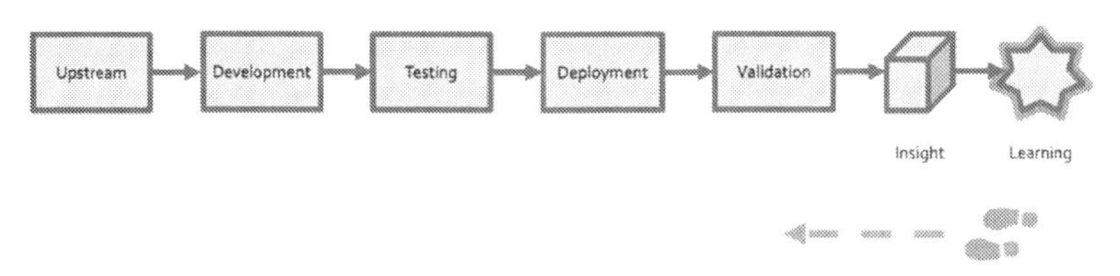

Figure 12. Upstream

I have used "user" and "customer" somewhat interchangeably, but it's important to keep in mind that the users of a product aren't necessarily the ones paying for it. Customers, sponsors, and other stakeholders are responsible for funding, goals, and various organisational policies and constraints, but are often poor judges of what really works for users. For Springboard DIY, retail customers spend money through the online site and they pay for the site itself only indirectly. That said, Springboard is under no illusions: only by delighting users will they keep coming back. This in turn means developers working closely with user researchers (part of the product team) and UX specialists (who at Springboard are regarded as having one in foot in the product camp and another in technology). In short, whenever developers don't have a real customer on hand to collaborate with, they have what they hope is the next best thing.

Feeding development are a range of activities that are so intertwined that any linear representation of them would be highly misleading. Here we see user research, ideation, analysis, prototyping, design, and so on – activities that once were managed separately but are now seen as just different facets of a process that has just one goal: to make sure that downstream activities (everything to their right) always have an appropriate amount of work of the highest possible value. Hence names like *Fuzzy front end*[32] and *Upstream* (upstream in the flow relative to development). Neither term is universally accepted; for convenience I'll use the latter, shorter one.

Upstream's focus on value goes a long way to explain its non-linear and multi-faceted nature. Looking all the way ahead to that final validation step, much hinges on the hypothesis and its assumptions about the user population, user needs, potential changes to user behaviour, technical challenges, and so on. Many of these assumptions can be tested long before fully-fledged solutions have been built, helping the organisation to learn very much faster and to maximise the impact of the work done downstream. And there's no better way to test assumptions about users than to actually go and talk to people, talking to users directly as well as to those that regularly interact with them. Customer groups, testing labs, fieldwork, and the customer helpdesk all have a part to play. Each of them presents different opportunities to test ideas, prototypes, and work at various stages of completion. Upstream has the potential therefore to involve people of every speciality, and the team at Springboard has a policy of ensuring that everyone does indeed get a regular turn.

Springboard has learned to stop thinking of upstream as a phase broken down into stages with gates in between; now they understand it as managing a portfolio of options for the best possible return. The best ideas will be released quickly, perhaps even before they are fully formed – the opportunity being great enough that more eyes and hands should be involved sooner. Some ideas take longer to mature, the cost/benefit equation being sufficiently marginal that a few rounds of prototyping and user testing might save the team some wasted effort later. The least promising ideas will languish for a while before a positive decision is made to reject them. No-one mourns a rejected option; each one is a bullet dodged, waste avoided.

Flow inefficiences

In place of the *7 wastes* of Lean production, here's a simple classification of the kinds of *impediments to flow* typically encountered in digital work:

- Impediments that impact individual *work items* – work items being the pieces of work that flow through the system – experiments, features, or small but still deliverable 'slices' thereof
- Impediments that impact the people doing the work
- Impediments that impact the broader system

Within these three broad categories are some more specific impediments. They're experienced everywhere, and most of their names are not only widely recognised within the Lean-Agile world but have a long history in the academic literature too.

When an individual work item is prevented from flowing, it is usually in one of two states:

- **Blocked** – A work item is said to be blocked when it is manifesting some abnormal condition that prevents it from progressing. The proximate cause is referred to as the *blocker*, and these too may be classified: unfixed defects, unresolved dependencies, missing information, relied-upon assumptions turning out to be invalid, and so on.
- **Stalled** – Not to be confused with being blocked, work items are stalled when there is no capacity in the system available to service them. Because the system's attention is elsewhere, this impediment often goes unnoticed.

The next two impediments impact most directly (and seriously) on people and systems:

- **Overburdening** – We saw this in the first chapter: people and systems overloaded and over-stressed, unable to perform at their best due to an excess of demand over capacity. On top of the human costs, overburdening has a devastating effect on quality and predictability, and Little's law still applies (try as you might, you can't beat the maths).

- **Starvation** – Just as people don't like to be overburdened, they don't like to be denied the opportunity to be engaged in meaningful work either. Starvation typically occurs downstream of a *bottleneck activity*[33] that is either behaving unpredictably (perhaps it is overburdened with too many work items or is choking on something oversized), or the bottleneck activity itself has been starved. Starvation is especially expensive when it occurs somewhere critical.

Before we move on to the last set of impediments, notice some important relationships between those we have covered already:

- Stalled work items and overburdening are two sides of the same coin, seen from two different perspectives. If you're seeing one, you are probably experiencing the other.

- Overburdening and starvation represent two opposite and undesirable extremes, indicating that something is out of balance. When either condition occurs system-wide, it usually means that demand has been managed ineffectively. When they seem to jump around between different parts of the system it likely indicates that there are improvements to be made in the direction of smoothness, for example dealing with blockers more proactively and learning how to prevent them, right-sizing work items, investing in automation, or cross-skilling.

Our final three impediments impact the broader system:

- **Defects** – Just as in manufacturing, defects in product development are double, triple, or quadruple whammies! Effort is expended first on producing defective work, then on detecting and rectifying the problem, and finally on dealing with any knock-on impact. The sooner they are found the better: so-called defects found locally by a developer running tests on new and still unshared code don't really count as defects at all – the failed tests are better regarded as placeholders for work that is still to be done. At the other extreme are defects that have "escaped to production", causing unwanted impact on customers that must be managed and perhaps even compensated for.

- **Failure demand** – A concept that originated in public sector service delivery[34], failure demand doesn't necessarily imply defects in the conventional sense that the work wasn't conducted as specified, rather that the work met customer needs so inadequately that the whole process must be gone through all over again, resulting in additional demand. On the surface, failure demand and failed experiments may appear to be similar, but experiments are framed to generate learning efficiently when outcomes are uncertain, whereas failure demand is mostly avoidable waste.

- **Unrealised opportunity**: This is the economic waste incurred through poor choices in the sizing and sequencing of work. Examples include high value work queued behind lower value work (perhaps because the latter has been committed to as part of a large batch), or imbalances between different types of work, sources of work, or *classes of service* (lacking the sophistication to balance date-sensitive work, urgency-driven work, and longer-term maintenance and improvement work, for example). No team has perfect information, but the best teams understand the economic basis of their work in general, have a good feel for the likely *cost of delay*[35] attributable to individual work items, and are capable therefore of making good decisions on the ground.

Taking these as a whole, it should be clear that there is far more to flow than just keeping people busy. "Busy-ness" may even be a bad sign! Lean-Agile means 1) managing flow for its customer value and 2) doing the right thing on behalf of the people in the system, both at the same time.

Managing flow, right to left

Without (yet) the help of certain tools and frameworks that will be introduced in the next chapter, a right-to-left strategy for managing flow reveals itself in the questions below. Keeping the value stream map in mind, they are posed in a right-to-left sequence:

- What are we learning from our working systems, both from changes recently deployed and from what has been in place for a while?
- What do we hope to learn soon, and are our systems and experiments set up to deliver the information we'll need?
- What do we have ready to deploy? When and how will it be deployed?
- What needs testing? How, by whom, in what environment, and against what criteria?
- What is in development? Who is working on what? What are we doing to clear our blockers?
- Are we clear about what comes next? (This last question can be left unasked if the capacity for new work won't be available soon)

These questions are easily learned – teams, managers, and coaches alike asking them out of habit. An incisive manager or coach focusses on the most pertinent questions; a mature team will have asked those questions of themselves already. When in doubt, start with questions that will help work to flow not just through the system but out of it. *"Stop starting and start finishing!"*

Improving flow, right to left

For even a single work item to flow through the system smoothly, a lot of things must go right! The questions above are designed to head off some of the most important ways that things can go wrong:

- Work has been deployed, but nothing learned, either because it's not being used, or because it is so inadequately instrumented that we lack the data we need
- We're ready to deploy but we don't know what we hope to learn (we have no clear hypothesis to test) or we won't generate adequate proof
- We have nothing ready to deploy now and won't have anything to deploy soon; alternatively, we do have things we'd like to deploy but are prevented for reasons of capability, availability, or authority
- We have nothing to test now and won't soon; alternatively, we're unready for some reason, or blocked by defects
- Work is blocked in development, work that perhaps should not have been started in the first place
- There is a shortage of valuable work in development or in the pipeline

How should a leader respond in these situations? A good response works at three levels:

1. Dealing with the immediate issue in hand, its impact, and its proximate causes
2. Dealing with systemic causes – the fact that the system allows these issues to arise
3. Dealing with the fact that this system deficiency wasn't anticipated before these issues were allowed to arise

Leadership responsibility increases with these levels. A team that is able to deal with its immediate issues and their proximate causes should be allowed to do so, and coached in that basic capability if it is not. Next, how might we improve the system to make similar problems less likely or less impactful? Lastly, why are we responding to this only now? Have we failed to look ahead, been too reactive, complacent even?

Blame the system, take responsibility

You've heard it before, no doubt: "Don't blame the person, blame the system!" W. Edwards Deming famously took this advice and calibrated it:

> *94% of the problems or defects in an organization are caused by "the system".*

Deming was speaking in statistician's code; in his mind the remaining 6% attributable to people was "almost insignificant". Sometimes he would raise the system's share of the responsibility to 95%; now the people's share was "statistically insignificant".

To help make this general advice more concrete and to temper any natural instincts to direct blame in unhelpful directions, try this default response to failure:

- Until demonstrated otherwise, most failures are failures of collaboration

In other words, had the right conversations had happened at the right time, or had people chosen to work together in a different way, failure would have been averted. Good processes encourage and reinforce these better behaviours until they become so well established as cultural norms that any deviations stand out and the system becomes naturally self-correcting.

Not every failure can be explained in this way (for example failures of foresight or investment, leadership responsibilities in both cases), but it's a safer default assumption than most of the alternatives.

Let me give a couple of personal examples:

1. Code reviews were once a frequent cause of long delays and intense frustration across the global department I once led. In every case, had the writer and the reviewer of the code in question collaborated appropriately before and during development, the code review would have held few surprises. Had they worked sufficiently closely (perhaps practicing *pair programming* as my next team did), the formal review step might not be necessary at all! Once we were able to recognise this problem as a failure of collaboration, improvement quickly followed.

2. A few years later in an interim delivery management role, I found that work would frequently arrive in test with no clear plan for what would happen next. Worse, when testing did eventually start, how long it would take would be anyone's guess. The remarkably effective fix: a quick *3 amigos conversation*, representatives of product, development, and test holding a short conversation about each piece of work immediately before its development starts. Not only did we head off those frustrating delays in test, we also set up the conditions for collaboration right across the process.

Changes like these are quick and easy to agree, don't require any great expertise, and their benefits vastly outweigh any costs. Of course not every problem is quite so easily dealt with – let alone prevented before it ever materialises – and some require significant investment. Fortunately, many problems of flow recur so frequently in the digital space that there is no shortage of well-documented ways to avoid them. In the next chapter, we'll cover some of the main frameworks and how they combine.

Reflections

1. How does your product development organisation promote collaboration, both internal (within and between teams) and external (with customers most especially)?
2. What would your product development organisation understand by the term *working software* (or more generally, *working product*)? How compatible is that understanding with the concept of *continuous delivery*?
3. How does your product development organisation adapt to changing knowledge – product-wise, process-wise, and organisationally? How does it stimulate and capture that knowledge?
4. How does your product development organisation maintain an appropriate balance of attention across delivery, development, and infrastructure (the last of those encompassing technology, process, and culture)?
5. Working from right to left, how do you understand your product development value stream?
6. To what extent is your product development value stream anchored on the right in validation? What forms does that validation take?
7. What "upstream" activities keep the product development process fed with high value work? How do the best ideas make it to the front of the queue?

8. How do you manage work out of (as opposed to into) your product development process?
9. How do you recognise, mitigate, and address these *"flow inefficiencies"*: blocked work; stalled work; people, teams, or systems either overburdened or starved of high value work; defects; failure demand; unrealised opportunity?
10. Working from right to left across your product development value stream as it is today, which of the above flow inefficiencies would you expect to encounter first? Were you to repeat the exercise after addressing that inefficiency, what would you expect to find?
11. How many of your recurring inefficiencies could be framed as *"failures of collaboration"*? How does that framing help?

---

[29] See also the blog post, *Incremental and iterative*, blog.agendashift.com/2018/02/22/incremental-and-iterative/ and the C2 wiki page, *History of Iterative and Incremental Development*, wiki.c2.com/?HistoryOfIterative

[30] Recommended further reading: *Continuous Delivery: Reliable Software Releases through Build, Test, and Deployment Automation*, Jez Humble & David Farley (2010, Addison Wesley)

[31] See Gall's law, en.wikipedia.org/wiki/John_Gall_(author)#Gall's_law, and John Gall's rather wonderful book *Systemantics: How Systems Work and Especially How They Fail* (Pocket Books, 1978)

[32] en.wikipedia.org/wiki/Front_end_innovation

[33] Technically, a process's *bottleneck activity* is the activity that constrains throughput, the rate at which work items can be delivered. See also *Theory of Constraints*, one of the frameworks introduced in chapter 3.

[34] *Freedom from Command and Control: A Better Way to Make the Work Work*, John Seddon (Vanguard Publishing Limited, 2003)

[35] *Principles of Product Development Flow: Second Generation Lean Product Development*, Donald G. Reinertsen (Celeritas Pub, 2009)

Chapter 3. Patterns and frameworks

Let's take stock for a moment. We have described:

- **Lean** as the strategic pursuit of flow, a process of organisational learning
- **Agile** as people collaborating over the rapid evolution of working software that is already beginning to meet needs
- **Lean-Agile** as the strategic pursuit of flow in complex environments, the organisation placing high value on collaboration, continuous delivery, adaptation, and learning

Whether or not you see equivalence in these informal definitions of Lean, Agile, and Lean-Agile, it is certainly true that their respective communities have identified or created some important ways to improve flow that are highly applicable to digital. And they're not mutually exclusive! Rather than seeing their documented solutions as alternatives – leaving you with the difficult problem of choosing between them – it's much more helpful to see them as complementary patterns that can be combined in interesting ways.

Here then are seven key patterns and the bodies of knowledge that best exemplify them:

1. **Iterated self-organisation around goals**, exemplified by **Scrum**
2. **Explicit attention to flow**, exemplified by **Kanban**
3. **People, process, and technology**, exemplified by **XP** and **DevOps**
4. **Exploring options**, exemplified by **User Story Mapping**, **Jobs to be Done**, and **BDD**
5. **Customer co-creation**, exemplified by **Service Design Thinking**
6. **Systematic bottleneck management**, exemplified by **Theory of Constraints**
7. **Hypothesis-driven business experimentation**, exemplified by

Lean Startup

For convenience, I'll refer to these diverse bodies of knowledge as 'frameworks', a word that has multiple meanings. Some of them – Scrum and Lean Startup most especially – are frameworks in the sense that they provide some minimal structure into which specific practices can be introduced. Others – DevOps and Design Thinking for example – are frameworks in the different sense that they provide a particular perspective to an organisational problem and an array of techniques with which to approach it.

Together with the scaling frameworks of the next chapter, the frameworks described here form some of the most important and easily-recognised features of the Lean-Agile landscape. Necessarily, this chapter includes a fair amount of detail, but even so, it only scratches the surface. For every topic mentioned, whole books have been written! As you see each framework introduced, bear in mind the following:

- If a framework seems to solve a set of problems that you just don't have, that's completely fine. Be thankful (and also watchful).
- If you come to the realisation that one of these frameworks has become the lens by which you view the others, experiment with taking an opposite perspective and notice how your understanding changes. A lot can be gained from putting models together, playing One Model to the Tune of Another, or throwing them into the Great Model Collider and seeing what flies off; innovation is often sparked this way.

The chapter finishes with a quick review, a reminder of how the right-to-left perspective and its relentless focus on needs and outcomes helps to unify all the frameworks covered.

Pattern 1. Iterated self-organisation around goals: Scrum

Scrum is a simple process framework that is well suited to product development, providing some basic structure into which specific practices can be introduced. I'll describe it in two contrasting ways, both of which are entirely compatible with Scrum's official description, the Scrum Guide™, which can be read at or downloaded from www.scrumguides.org.

First, let me introduce Scrum as *"iterated self-organisation around goals"*. That's my phrase, and it summarises a right-to-left description:

- A Scrum Team moves towards its objectives goal by goal.

- For a timeboxed interval called the Sprint, the team collaborates around a shared goal. At the end of that period, the team reflects on how well the Sprint Goal was achieved, looking for ways to improve. It then prepares for the next Sprint, an opportunity to try new ways of working as it organises itself around a new goal.
- The team's best understanding of the work required to achieve the current Sprint Goal is represented by its Sprint Backlog (practically speaking a list of work items); options for future Sprints are maintained in a Product Backlog, an expression of unfolding strategy and the responsibility of the Product Owner.

If you're completely new to Scrum, welcome! If on the other hand you're already familiar with it, you may be thinking that this right-to-left description of Scrum is quite unlike any you have read before.

Here's a much more conventional description, one that starts from the left:

- The process starts a with Product Backlog, for which the Product Owner is ultimately accountable
- At the Sprint Planning event at the start of each Sprint, the team creates a Sprint Backlog, selecting work from the Product Backlog and deciding how best to implement it
- Meeting each day for the Daily Scrum, the team does its best to complete the work identified in the Sprint Backlog during the Sprint (a fixed period, typically two weeks in duration)
- The Sprint finishes with a Sprint Review and Sprint Retrospective, in which (respectively) the output of the Sprint – a Potentially Shippable Increment – is reviewed by all interested parties and opportunities for learning are captured within the team

So which of these two very different descriptions is correct? As I have already stated, both are entirely compatible with the Scrum Guide. However, the first, right-to-left version of Scrum far better describes the skeleton of a process for "*developing, delivering, and sustaining complex products*" (to quote the first line of the Scrum Guide) than does the second, left-to-right version. The latter could easily describe a linear, plan-driven process that provides little opportunity for the product and its environment to evolve together in any meaningful (and value-creating) way.

You might be tempted to think that what separates the two versions is *pull*, but be careful: technically, both have a pull system at their heart. Regardless of whether it is expressed in terms of work items or goals, no more than a Sprint's worth of work is selected for each Sprint Backlog, the rest remaining uncommitted in the Product Backlog. However, the very real risk with left-to-right Scrum is that Product Backlog is treated like a detailed plan, with too many up-front commitments made or assumed, and a push mentality begins to dominate the system as a whole. Allowed to go unchecked, over-committed Sprints lead to overburdened teams, and as we know, overburdening leads to number of serious consequences.

Even if not over-committed, fully-committed Sprints leave little space for experimentation or for feedback to be acted upon, and it is this as much as any technical aspect of process design that compromises agility. The obvious countermeasure is to factor in the time to collect and act on the evidence that meaningful goals are actually being achieved, something the right-to-left Scrum team does by default.

I hope that I have made the contrast between the two kinds of Scrum sufficiently clear. It's a matter of regret that so many introductions to Scrum miss the opportunity to take a more right-to-left stance, with material consequences for how it is adopted and operated. Turning that around, decisions surrounding how you adopt a framework such as Scrum and make it work for you in your context can matter at least much as your initial choice to adopt it. We'll return to this important issue in the next chapter.

Scrum and Leadership

Given both the popularity of Scrum and its strong emphasis on self-organisation, it's important to understand the role of leadership, both from within the Scrum structure and from outside. For Scrum to succeed both on its own terms and organisationally, four critical things need to be present:

1. **A clear sense of purpose**, which for the product dimension is embodied in the Scrum role of the Product Owner. Other dimensions of purpose and intent, such as architectural goals and the organisation's overall mission aren't directly addressed in Scrum; they must be dealt with by the team through its relationships with the organisation – whether directly, via the Product Owner, or via the Scrum Master, whose job is to help make Scrum work effectively.

   As purpose unfolds into goals and from those into action, the ability to help articulate and explore outcomes is a key leadership skill, especially where self-organisation and innovation are to be encouraged. For Scrum specifically, the Sprint Goal is of course a kind of outcome, an outcome that gives purpose to a Sprint. Working backwards, it's important to express longer-term plans such as product and technology roadmaps in the same kind of terms, avoiding unnecessary detail and premature specification. This creates the space for creative collaboration to happen at the right time.[36]

2. **The right size of team, with adequate self-sufficiency**, given the team's purpose. These two requirements are somewhat in conflict; few teams are entirely self-sufficient, and the smaller the team (Scrum recommends teams of between 3 and 9 people), the less likely it is that it has a complete set of skills at its immediate disposal, increasing its dependency on others outside the team. A good self-organising team will recognise and address gaps or vulnerabilities in its collective skillset to create a *cross-functional team*, but this takes time; meanwhile they must learn to manage its external dependencies effectively. Clearly, good choices around staffing and the right investments in skills can pay dividends.
3. **Trust in the process.** If your organisational culture is underpinned by assumptions like *'leader knows best'*, and *'leaders tell people what to do'*, the transition to self-organisation can be a difficult one. Fortunately, any leader who is serious about change can easily test some alternative assumptions, such as '*the people closest to the work generally know the most*', and *'leaders give space for people to give their best'* (noting that space is designed into Scrum in the form of the Sprint). Allowing some time for trust to be built through delivery, new assumptions underpin new values, and culture begins to change.
4. **Engaged governance**, synchronised to Scrum's rhythms and helping both team and organisation support each other in their continued evolution. Not 'drive-by governance' (where participants with no skin in the game feel their job is done once they have found an issue for someone else to address) or mere 'issue escalation' (in which ever-growing lists of issues are collected and only sometimes resolved), but opportunities for the internal contradictions of a changing organisation to be surfaced and owned by the right people. We'll keep returning to this important theme in later chapters.

It would be fair to say that this list identifies some key elements of good leadership and good organisational design, with or without Scrum. Experience however teaches us that Scrum has a habit of bringing issues of clarity, scope, trust, and governance into sharp and sometimes painful relief. That presents a challenge, but also the kind of opportunity that good leaders wouldn't want to waste.

Pattern 2. Explicit attention to flow: Kanban

In the factory floor kanban systems of the kind described in chapter 1, signal cards (the *kanban*) are sent upstream just soon enough for finished goods, subassemblies, or materials to arrive as needed, just in time, inventory kept to a minimum. A simple but surprisingly effective right-to-left signalling system results in smooth, left-to-right flow.

Away from the factory floor, kanban systems often look very different, typically taking the form of sticky notes or cards that move across some kind of whiteboard, wall, or electronic screen – the *kanban board*. The intent however is much the same: explicit and continuous attention to flow[37], both for its own sake and for broader impact it can have on organisation design in the longer term.

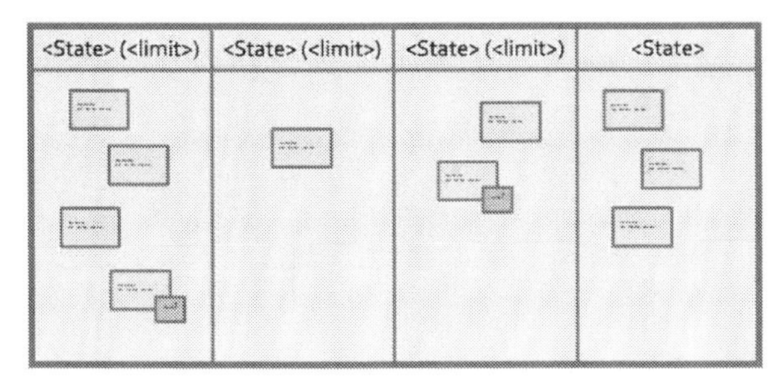

Figure 13. A simple kanban board

This new kind of kanban system has three key elements:

1. A visual representation of the workflow at some appropriate level of detail. Typically, this takes the form of columns drawn on some kind of board (a whiteboard or an electronic tool), arranged in left-to-right sequence, the heading of each column being with the name of a work item state or its corresponding activity[38].

2. Visual representations of individual work items, each described at a level of abstraction and size appropriate to the workflow, and positioned according to where in the workflow they currently reside. For the kind of board described above, each work item is represented by a sticky note or card which is moved across the board from left to right as the work progresses towards completion. Additional state information – the presence of blockers, for example – can be represented with other visual cues.

3. Policies designed to control the number of cards in the system or parts thereof, thereby controlling the amount of work in progress (WIP). The most important of these policies are *work-in-progress limits* (WIP limits), simple numbers that govern the maximum number of work items allowed in affected columns (or more generally, in defined regions of the board).

With all three elements in place, we have a pull system: work gets pulled into the system (or parts thereof) only as capacity allows. At a stroke, overburdening is alleviated, and the reduction in WIP increases the opportunity for collaboration. Many of the teams that I have worked with in recent years keep fewer work items in progress than they have people, suggesting that there must some collaboration happening most of the time (and there is)[39].

In one very practical sense, the kanban board is the very embodiment of right-to-left thinking. Here's how we typically review the board, in a standup meeting for example:

- Starting at the right hand side of the board with work recently completed, will it stay completed? What did we learn from completing it? What are we still learning, now that it's in use?
- What can we do to get our nearly-completed work over the line?
- Are there any issues blocking our in-progress work? What's being done to unblock them? Is everything else progressing as expected?
- Do we have the capacity to start new work? If that opportunity is coming soon, do we understand which work will be selected next, and why?

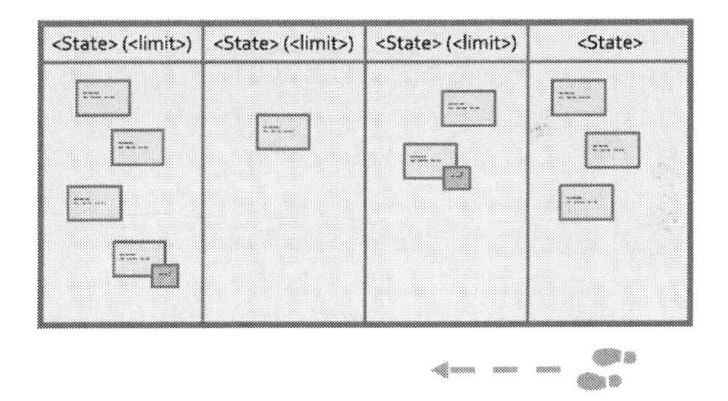

Figure 14. The board is reviewed from right to left

From this right-to-left style of day-to-day work management, it's only a short step to a right-to-left style of process improvement. With very little prompting (or rather, the kanban board is the prompt), managers and teams alike learn quickly to recognise and address anything that seems to impede work getting out of the system. Whenever the impediment is some aspect covered by the visualisation or policies of the kanban system, the design is altered and the process changes at a stroke! Eventually, as work begins to flow through the system more smoothly, any questions over the quality of the work coming into the system increase in importance, and attention turns upstream.

The Kanban Method (as codified by David J. Anderson and described in his 2011 "blue book"[40] and later given a values-based treatment in my first book[41]) very clearly describes not a development process but an improvement approach. It is characterised by a deliberate avoidance of prescription around implementation detail, and has a very Lean-like pursuit of flow at its heart. Compared with the traditional Lean approach as described in chapter 1 there are however a couple of noteworthy differences of emphasis:

1. It is an evolutionary, *"start with what you do now"* approach – advising against wholesale changes until the existing system has been stabilised through visual management and controls on WIP, after which the system should be much more amenable to change.
2. It makes little mention of waste, focussing instead on the wide array of positive steps that can be taken: visualisation and WIP control most obviously, but also the design of feedback loops, the use of metrics, and the practice of hypothesis-based change (more on that last one later in the chapter).

Scrum and Kanban

Kanban from the Inside came out in 2014. I mention the year, because back then I felt that I was going out on a limb by including in a Kanban book a chapter favourable towards Scrum. Now, just a few years later, the era of the "Scrum versus Kanban" blog post isn't quite dead, but when such articles do come out, they are usually dismissed pretty quickly, and from both sides. In truth, the only real competition between them is for training dollars; it takes only a little understanding of the two frameworks to understand that they are more complementary than conflicting. In fact, the Scrum and Kanban combination (sometimes called Scrumban) has a lot to recommend it.

It's not hard to see that they should be complementary:

- Scrum describes a team-level delivery process highly suited to product development; Kanban works with a wide range of existing processes at a range of scales – e.g. personal, team, value stream, and project portfolio – making it capable of working both inside the Sprint and around it
- Scrum defines roles (Product Owner, Scrum Master, Team); Kanban does not
- Scrum's Sprint Backlog provides a ceiling on work in progress and

(with the Sprint Goal) a clear focus for the Sprint; Kanban allows a range of controls to be applied to whatever parts of the process will most benefit

- In complementary ways, both are *outcome-oriented* – in Scrum through the Sprint Goal and all the upstream activities that contribute to identifying that, and in Kanban through the longer-term pursuit of *fitness for purpose*

In practical terms, there is no single Scrum and Kanban combination; instead they combine in a number of ways. Listed roughly in order of increasing interaction between the two (or in other words, least interesting first), here are four such combinations:

1. Scrum provides the core development process; Kanban picks up everything that doesn't comfortably fit – support work and urgent enhancements, for example[42]
2. Kanban is used to manage the Scrum Team's tasks – eg development tasks or testing tasks – using a simple "To do / Doing / Done" structure or a refinement that better describes the team's typical process
3. Because *"you can't deliver a task"* (to coin a phrase), Kanban is used to manage features from the start of the Sprint (or perhaps a little upstream) through to completion, a moment that may or may not coincide with the end of the Sprint[43]
4. Perhaps in combination with one or more of the above, Kanban is used outside and around the Scrum process to manage defined themes or batches of work – business initiatives, longer-running product experiments, *epics*[44], etc – out of which feature ideas are generated and tested

It should be clear now that Scrum and Kanban are fulfilling two quite different roles: Scrum the beating heart of the development process, Kanban a powerful coordination mechanism both inside the Scrum process and across Sprint and organisational boundaries. When given the chance, they combine to promote flow, self-organisation, adaptation, and learning.

Pattern 3. People, process, and technology: XP and DevOps

More than a decade separates Extreme Programming (XP) and DevOps, but they have much in common. XP predates the Agile manifesto and remains the inspiration for many of the technical practices followed by development teams today. DevOps set out to address an organisational problem, but in so doing became the catalyst of rapid advancement in both process and technology. Together, they provide the technical underpinnings of the modern software development process, especially for digital, where speed and feedback are at a premium.

XP

In an influential 2009 article titled *Flaccid Scrum*[45], Agile Manifesto signatory Martin Fowler used words like "mess" and "crippling" to describe what happens when teams adopt Scrum (regarded at the time as the *de facto* Agile process) without also adopting the kind of technical practices brought to the fore in XP. These teams were finding that their productivity was plummeting because their codebases were becoming increasingly unmaintainable. So much for the manifesto's commitment to *working software*!

Nowadays, when proficient development teams refer to Scrum, they often mean Scrum + XP, or at least Scrum plus many of XP's technical practices. This means:

- The team's codebase advancing in small, tested increments, automated tests being developed by the developers themselves in parallel with (and even slightly in advance of) their functional code
- The changes of all team members being integrated many times a day, each integration automatically triggering a complete run of the test suite
- Developers reviewing each other's work, either as a standalone event or (better) through pair programming
- Regularly *refactoring* code to improve its maintainability, the test suite ensuring that the new code remains functionally equivalent to the old

None of these practices should be considered separate tasks or overheads; quite simply they reflect (even now, 20 or more years later) our best understanding of how teams of people can develop software in a sustainable way. You could say that failing to test, integrate, review, or refactor continuously is storing up trouble for the future; XP cleverly flips this around with the mantra *"If it hurts, do it sooner and more often"*.

To do this effectively requires a mutually-reinforcing blend of people, process, and technology:

- **People** committed to sustainable development, which implies commitments to quality and to skills. Many of these skills are embodied in what has come to be known as *Software Craftmanship*[46] but it's worth noting that some of these skills – the ability to identify small increments ('slices') of testable work, for example – are of great value to the broader process and can be learned by non-programmers.

- A supportive **process** that amplifies feedback, builds quality in, and encourages knowledge to be shared quickly and widely.
- **Technology**, including as a bare minimum:
    - Shared code repositories into which all changes are committed
    - Testing frameworks – code libraries that make it easy to write automated tests in an idiomatic style that any developer on the team will readily understand, often to facilitate *test driven development* (TDD)
    - The *continuous integration* (CI) systems that merge the work of developers and run test suites and other automated quality checks

The sooner that these are established, the easier it is and the greater the benefit. The tools are so ubiquitous now – available on an open source basis on every major platform – that the only real barriers to entry are organisational. And consider this: wouldn't it be more than a little embarrassing if your technology team suffers for the lack of technologies that are essentially free?

The *"sooner and more often"* thinking of XP has its counterparts outside development, most visibly in the quality assurance (QA) community and in the DevOps movement. In the case of QA, the value of *"Test early and test often"* has been understood for a long time. Parts of the QA community take this further, encouraging responsibility for quality to *"Shift left"* (their phrase, not mine) and the role of QA to change from a last-ditch activity to something that works both with and on the whole process.

DevOps

Little more than a decade old, DevOps is much more recent than XP. Not so much a process as a movement, its goal is to drive the integration of software development and IT operations. The benefits of dismantling this silo barrier are clear:

- Through meaningful cross-functional collaboration, software is designed for supportability, with benefits both to the people doing the support (regardless of which department they happen to belong to) and to their customers

- Continuous Integration (as described above in relation to XP) is extended to *Continuous Delivery* (CD), increasing the scope of automation, greatly increasing the rate at which deployments can be made and substantially reducing their risk
- Fast feedback around operational concerns such as availability, performance, capacity, security, data retention, maintenance cycles, technology dependencies, and so on

In this supposed era of the cross-functional team it is perhaps a little embarrassing that Agile should need DevOps. But to be fair, it has long been the practice in many organisations to keep development and operations in separate silos, a legacy that will be overcome only with prompting and support. And to its credit, the existence of DevOps has created a focal point for rapid technical innovation that continues to benefit even the most integrated of digital organisations.

As with the tools associated with the XP practices, many of the technologies of DevOps are freely available. Some of the best-known of these have been open-sourced by Google, Netflix, and other digital giants that operate IT infrastructures at scales never previously seen. But it's not just about scale; what stands out is the sheer rate of delivery. Even with the scalable web servers, shared web services, microservices, and serverless functions of the modern architecture, it takes a large number of application deployments to achieve the millions of updates per year made by some of these companies.

Suddenly, production engineering has become exciting, and the improvements are felt upstream. Not only can deployments be done at will, but thanks to the virtualisation of servers and networks, whole environments can be created, destroyed, scaled up, or shrunk down as needed, even temporarily as part of the CI build process. And thanks to containerisation, services can be built once and then run without modification in every environment, whether on a developer's laptop, a test server, or in production.

Pattern 4. Exploring options: User Story Mapping, Jobs to be Done, and BDD

Before we continue upstream to pattern 5 and Service Design Thinking, it's worth pausing to introduce some of the product management tools most visible at development's upstream boundary.

A popular and effective tool for organising features for later implementation is the *Story Map*, short for *User Story Map*[47]. Visually, a story map is easily confused with a kanban board but instead of the column names describing the stages of a delivery process or the states of work items, the headings – the story map's *spine* – together describe a high-level *user journey* through the system under construction. Under these headings, *"stories"* (more about these in a moment) are prioritised vertically; when the time comes to select work for development, items are taken from the top of one or more columns chosen according to the team's current focus.

To give a simple but real example, when I was interim delivery manager for the *Find an apprenticeship* digital service, we had a 5-column story map with these headings:

1. **Search** (for available apprenticeship vacancies)
2. **Apply** (for a particular vacancy)
3. **Manage my applications** (track the status of applications, withdraw applications, etc)
4. **Manage my account** (amend candidate profile, change password, delete account, etc)
5. **Support candidate** (features for external careers advisors or internal operations staff)

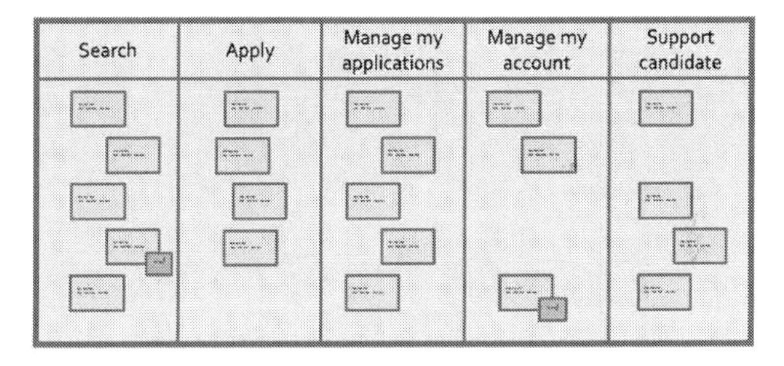

Figure 15. A user story map

During my involvement in the early stages of the service's development, most user-facing features and enhancements would find a home in one of those five columns. Later, as the candidate-facing service matured and more development capacity could be allocated to the needs of employers and training providers, the story map would evolve to suit. Among those evolutions was the visualisation of dependencies and other actual or potential blockers; encouraging us to manage them proactively and to prioritise the affected features carefully. Don't start what you can't finish!

The *user stories* that give the story map its name briefly describe features or slices thereof. In the words of Kent Beck, their inventor, user stories are just *"placeholders for a conversation"*. Often those conversations start in a deliberately stereotyped way, conforming to a user story template such as this one, known as the Connextra template in honour of its source:

- As a <*persona*>, I want <*goal/ desire*> so that <*benefit*>

Unfortunately, it's not uncommon to see user stories that don't really describe user needs. It's as through user stories have been reduced to this:

- As the product team, we want <*requirement*> so that <*business justification*>

And we know what happens when we start ploughing through lists of requirements: mediocre products!

One important technique to avoid this tendency is to bring features to life through the *scenarios* – the *"authentic situations of need"* as I like to describe them – in which the features become important to someone. As leaders, you can just ask:

- When is this important? Can you describe an authentic situation of need for this feature?

Where it would be useful to capture the answer, consider the short-form format of the *job story* and long-form format of Behaviour Driven Development (BDD).

The job story template looks like this:

- When <*authentic situation of need*>, I want to <*action*>
  so I can <*achieve outcome*>

Job stories come from the *Jobs to be done*[48] (JTBD) school. One of their classic examples is borrowed from Harvard Business School professor Theodore Levitt:

> *"People do not want a quarter-inch drill, they want a quarter inch hole"*

Now you know what inspired the story that opens chapter 1! Here in one sentence is the *output* of that trip to the DIY store – the drill – and its *outcome*, the hole.

BDD is a specification language that emerged from test-driven development (see XP above) and has gained popularity with business analysts and testers too. In BDD, user stories are augmented by multiple scenarios described in a three-part Given/When/Then structure, the Given part capturing the scenario's initial conditions, the When part identifying a triggering event that provokes a response from the system, and the Then part describing the expected results (with an outcome achieved and a need met).

For example:

> **Story**: Location-based search
>
> **As a** candidate, **I want** to see vacancies near me **so that** I can keep my travel costs low
>
> **Scenario**: Candidate provides a valid postcode
> **Given** a valid postcode
> **When** the candidate searches for vacancies
> **Then** results should be returned closest-first
>
> **Scenario**: Candidate provides an invalid postcode
>
> *etc*

A number of automated testing tools support BDD-style specifications, making tests capable of being read (and to a limited extent written) by non-programmers.

It's worth stressing that regardless of the practices and tools you employ, specification – like everything in a well-run continuous process – is a just-in-time activity. Here, this means:

- **Not too early** – being careful to avoid prematurely specifying things that are subject to change, based on untested assumptions, or stand little chance getting to the front of the queue for development anytime soon
- **Not too late** – maximising opportunity and minimising later pain through anticipation, proactivity, curiosity, imagination, and empathy for the customer
- **Not too little** – bringing underlying needs and hoped-for outcomes to the forefront; identifying key risks, dependencies, and edge cases
- **Not too much** – taking care not to over-specify detail that could be left for later collaboration; sizing work items appropriately – breaking down larger work items and sequencing the slices; refraining from working on specification when that effort would be better spent elsewhere

These considerations are all highly contextual, depending not only on the individual work items in question and the opportunities they each represent but also on the capabilities and current capacity of the team as a whole. Given that specification can be an easy scapegoat when anything goes wrong downstream, don't underestimate the experience, skill, and personal qualities required to maintain an appropriate balance over time.

Perhaps the most liberating aspect of doing specification just-in-time is that work items whose time hasn't yet come feel much less like requirements and much more like options, to be managed on an individual and portfolio basis. Generating options at the right rate is a cheap way to promote innovation, adaptability, and resilience. Developing and exercising options at the right time is how outcomes are maximised overall. Allowing less valuable options to lapse rather than adding to inventory is just good housekeeping.

Pattern 5. Customer co-creation: Service Design Thinking

Moving upstream, Service Design Thinking brings together these two disciplines:

1. *Service Design* – making services easier and more desirable to access, use, and support
2. *Design Thinking* – a multi-disciplinary approach to creative problem-solving

How do you know that a BDD scenario really does describe an authentic situation of need? Or that a user story's persona is usefully representative? There's really only one way to be sure, and that's to go and see. This is one way in which Service Design Thinking can get seriously multi-disciplinary, with the potential to involve user researchers, and even anthropologists or ethnographers (to understand potential users in their social and cultural context, for example). Not that the presence of such specialists should exclude technologists or operations staff from fieldwork – in fact the reverse! From first-hand experience and from seeing team members rotated through user research, I know that the more that team members are included, the greater the empathy they have for their users and the more motivated they are to find good solutions to the problems they witness.

These examples from the *Find an Apprenticeship* service illustrate the importance of context:

- The 16-year old who tells us *"But I wouldn't be doing this on my computer, I'd have my phone out during lesson time, below my desk, out of sight of the teacher!"*
- The 18-year old who is considering an apprenticeship instead of a university degree (that these would be considered alternatives is a relatively recent phenomenon in the UK)

- The candidate that would consider "anything within a bus ride" (emphasising to us the importance of location and raising the potential for more sophisticated integrations with online maps)

Only through fieldwork and with plentiful examples can truly representative personas be constructed. Often, it's the *when* and *where* of real situations that's the key to unlocking the *who*, *what*, and *why* of the user story.

Having identified some authentic needs, they must be prioritised. Some needs are especially important: they help to define what the product or service is all about. As we'll see in chapter 5, some might even be described as 'strategic needs' – needs that help to define the organisation's mission and give shape to products and services. For example, bringing candidates and employers together for apprenticeships is a strategic need fulfilled by the Skills Funding Agency, as the relevant UK government agency was then called[49]. It isn't its sole function but it is certainly a key part of its mission.

In common with many other facilitated processes, Design Thinking is characterised by alternating phases of *divergence* and *convergence*. Divergence is the result of generative activities – both field-based and workshop-based – that create lots of detail (often captured on sticky notes) and wide ranges of alternatives: different needs, alternative solution ideas, competing assumptions, and so on. Convergence is a process of narrowing down, through clustering, prioritisation, testing, or other means.

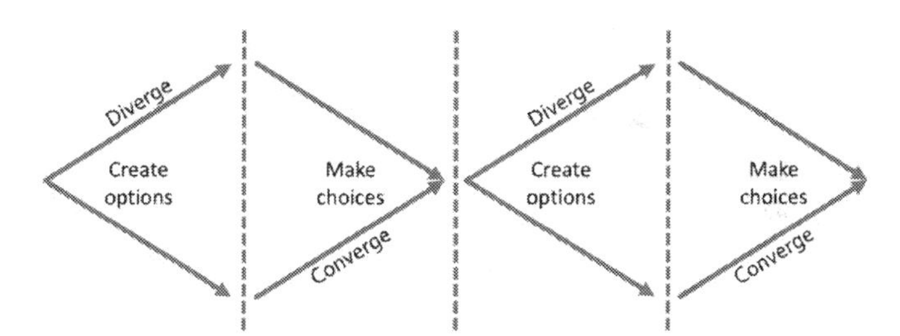

Figure 16. Divergent and convergent phases

What sets Design Thinking apart isn't just the use of tools familiar to designers, but the degree of user involvement in many of its activities. Users just aren't passive subjects of research, but active participants in a process of co-creation. In Service Design Thinking in particular, 'user' means anyone who interacts in any way with the service in question – customers, frontline and back office staff, managers, suppliers, and so on.

'Testing' too has a special meaning in Service Design Thinking. We're trying out our ideas on real people, and testing them through conversations, visualisations (movie-style storyboards, for example), paper prototypes, screen prototypes and so on. By not jumping straight to fully fleshed-out technical solutions, iterations can be completed remarkably quickly (and I've witnessed multiple design iterations completed in a single evening). Again, multiple disciplines come into play: service design, interaction design, user experience (UX) design, visual design, content design, and so on. The range of different perspectives combine to enhance the desirability of the service and its overall quality as experienced both externally and internally.

Crucial to the success of any new product is ensuring that the *whole product*[50] has been properly explored, not just the various (and likely contradictory) conceptions of the *core product* that its various stakeholders have in mind at the start of the process. Given that almost any interesting product has a service component, Service Design Thinking has much to offer even in traditional product development. For digital, it seems essential.

Service Design Thinking and Digital Leadership

Marc Stickdorn and Jakob Schneider's book *This is Service Design Thinking*[51] introduces five principles for effective service design, which they summarise as "user centred, co-creative, sequencing, evidencing, and holistic". To conclude this section, allow me to re-frame these principles slightly as acts of digital leadership:

1. **Start with (user) needs**[52]. If it's not meeting needs, what's the point? This might seem obvious, but do not underestimate the tendency for teams, managers, and sponsors to believe their own propaganda, to fixate prematurely on inadequate solutions, and to consider problems solved long before any evidence to that effect has been observed. The authentic voice of the user must be heard often enough and widely enough that respect and empathy for users and their needs drives the delivery of increasingly effective products and services.

2. **Bring people together**. Different stakeholders won't all need the same things – in fact it is almost inevitable that there will be tension between competing needs. Treat it not as a zero-sum game but as an opportunity for creative collaboration, involving stakeholders in the design process. When you bring together people with different perspectives, the chances that previously-unconsidered options will emerge are greatly enhanced, and everyone involved feels that they have a stake in the solution.

3. **Make it visual**. When there are a lot of moving parts, and especially when different parts of the organisation are involved, it's important that everyone understands how the service actually works. Visualisations such as user journey maps and service blueprints really

help to bring it to life, showing the sequence of events, the actions, and the interactions that combine to achieve desired results for all concerned. New designs can be tested on paper (literally!); with skilled facilitation and the right people in the room, visual design tools can enable service designs to progress rapidly, building shared understanding and confidence in the process. Afterwards, you can put the generated artefacts on display around your workplace, not just for reference or decoration but focal points for ongoing conversations.

4. **Make it tangible**. By what means will your users recognise that you have a valuable service to offer? How can they be confident that good service has indeed happened, even when most of the work is done behind the scenes? Between those two points, how will they know what options are available to them? If a service is to be engaging, it must be designed to reveal itself at the right times, at the right level of detail, and with the right visual impact, making something intangible – the service – somehow tangible to the customer.

   One important way to make virtual things tangible is through metaphor – 'store', 'marketplace', 'library', 'helpdesk', or 'forum', for example. Be aware that metaphors work at a subconscious level and create strong expectations of how the system will work, so choose them carefully and take care to respect their idioms.

   Sometimes, concepts as fundamental as 'user' can seem rather abstract to people working inside the service. Here too, look to make them more tangible. Give names to your personas, tell their stories (real, composite, or imagined), visualise!

5. **Look beyond the system**. From a design perspective, an engaging and effective service must work well at the boundaries between the digital and physical worlds. For example, one way that online retailers can reduce user frustration is to integrate with their distribution networks so that users can track their purchases through to delivery without being forced to sign into multiple systems, enter tracking codes, and so on.

From a systems perspective, it's also important to keep exploring the mutual impact (positive and negative) of the service and its business environment, paying particular attention to the ways in which key relationships might evolve over time. Among the more obvious things to watch are external competitive pressures and the internal appetite for ongoing investment; less obvious perhaps are things like hiring and staff development (moving targets as services mature), internal competition, synergies, and the regulatory environment. Not everything can be controlled, but don't underestimate your influence!

Pattern 6. Systematic bottleneck management: Theory of Constraints

The management framework *Theory of Constraints* (TOC) came to the world's attention in 1984 in the unusual form of a best-selling and still highly recommended business novel, *The Goal*[53], written by TOC's creator, Eliyahu (Eli) Goldratt. In common with Lean, TOC has its roots in manufacturing, has much to say about improvement, and has wide applicability outside its original field. The framework and the book both remain highly influential, a fact acknowledged in the DevOps movement's breakout book, *The Phoenix Project*[54], another business novel.

At the heart of TOC is an improvement cycle called the *Process of Ongoing Improvement* (POOGI) and its *5 focussing steps*. Here's a modern take on that model as described in Clarke Ching's brilliant short book, *The Bottleneck Rules*[55], based on a mnemonic, FOCCCUS.

FOCCCUS stands for **Find**, **Optimise**, **Coordinate**, **Collaborate**, **Curate**, **Upgrade**, and **Start again (strategically)**. If that seems a lot to remember, think of it as 'FOCUS', with the 'C' broken into three, like this:

1. First, we must **Find** the bottleneck, the activity that determines the overall process's rate of delivery. Typically, this is the activity that has the most work queued up immediately behind it or the one that has starved activities waiting on it; both conditions can apply at the same time.
2. Then we **Optimise** the bottleneck, squeezing what extra performance we can from it (we're talking quick wins at this stage, not fundamental changes).

Then the 3 C's – Coordinate, Collaborate, and Curate:

3. **Coordinate** between activities outside the bottleneck, removing distractions and adjusting the timing or pace of surrounding activities to suit the bottleneck better.
4. **Collaborate**, everyone looking for ways to improve the process for the benefit of the bottleneck. Almost by definition, this must also be for the benefit of the process as a whole.
5. **Curate**, ensuring that we get not just the *most* from the bottleneck activity, but the *best*. At all times it should be clear which are the most important items for the bottleneck activity to work on. Often it's a

good idea to maintain a small *buffer* of high value items maintained immediately upstream of the bottleneck so that it never runs out.

Only after the 3 C's do we:

6. **Upgrade** the bottleneck, investing in it to increase its capacity. Bear in mind that thanks to the improvements made in the preceding steps, the apparent need for significant investment may have gone away.
7. **Start again, (strategically)**, going back to the **Find** step. It's highly likely that the bottleneck has moved, and if it's not where we want it (a strategic choice) or if further improvement is still required, we repeat the steps.

Whether we're using the classic model or FOCCCUS, experts know to start with a crucial step 0:

0. Define the system's goal or objective

Step 0 is important because teams behave differently according to how they understand their objectives. For example, a team optimised for production behaves differently to a team optimised for service. A team optimised to write code behaves differently to one optimised to meet user needs. A team that is unclear about its purpose might be forgiven for optimising for its own comfort or for compliance to some arbitrary set of rules.

Continuous delivery demands continuous discovery

TOC warns us that bottlenecks move. In digital delivery, their movements tend to follow two predictable patterns:

1. In the absence of a good deployment capability, significant time may pass before it becomes painfully apparent that this deficiency is in fact a critical bottleneck. Only after addressing the problem is it realised just how much opportunity was lost forever by not having it in place sooner, because the ability to deliver continuously fundamentally changes the way people work.[56]
2. After surprisingly few improvement cycles, the key bottleneck often isn't anything technical but the lack of high-quality ideas in the pipeline, often accompanied by frustrating delays in getting decisions made. The problem isn't the time spent on analysis (any shortfall here can often be made up for downstream) but the lack of effort put into discovering and developing opportunities.

'Projectised' organisations typically learn these lessons painfully slowly. This is only partly explained by the fact that they do everything in bigger batches that take orders of magnitude longer to process. That's bad enough, but it's also because:

1. They tend to defer investment in deployment-related and production-related capabilities until the last responsible moment (or later), denying themselves the opportunity to practice these new high-feedback ways of working. When the ironically-named 'lessons learned' meeting finally arrives, it is already far too late.
2. Once the scope of a project has been decided, the mere suggestion that there might be more needs to discover and respond to is often actively discouraged. If discovery happens at all, it is done by people outside the delivery team in preparation for future projects, again limiting the opportunity to integrate new learning into current work.

Something as long-lived, open-ended, and business-critical as a core digital product or service demands a change of thinking, away from time-bound projects and towards more continuous models of delivery. My favourite example of an organisation facing up to this challenge isn't a tech giant, retail brand, or 'unicorn' startup, but the UK's Government Digital Service (GDS). They decided that no new government-sponsored digital service would be allowed to proceed past key checkpoints until it can demonstrate:

1. The level and availability of technical capability across development and operations needed to support fast and cost-effective service evolution, without compromising on the high standards required of government-owned systems in areas such as capacity, reliability, and security
2. A credible plan to sustain this service evolution in a highly user-centric way and well into the future, with user research and user testing established as core capabilities of every service team

In that government setting, these tests represented a radical change of strategy. And instead of waiting for each government service to stumble into these inevitable bottlenecks, foreknowledge of these tests incentivised service teams to address them proactively in whatever manner made most sense in their own particular context. If cash-strapped public sector organisations could make these bold changes in times of austerity, what excuse does your organisation have?

Pattern 7. Hypothesis-driven business experimentation: Lean Startup

"Everything I've seen in the data since that release looks great so far" said Rowan at the start of the previous chapter. What kind of data would the product manager

be looking at? To answer that question, it is necessary to go a few days further back in time. Alex, the developer involved in that conversation, is sharing an idea.

Alex:	So I was walking past the DIY project leaflets near the store exit and I wondered "What if we had these online too?", and that got me thinking. I even stopped a couple of customers to ask if they would use it! I know that you can't believe everything that people say in interviews, but even so I think it could be great!
Rowan:	I can see that you're excited! Suppose we really went to town on this, what would it look like?
Alex:	Well, we'd get all those leaflets pages converted into web pages and we'd make it really easy for the customer buy everything they'd need – all the bits and pieces and the more expensive stuff like tools too.
Rowan:	Cool! And we'd make sure those project pages got some decent SEO…
Alex:	Exactly – and we'd add links to them at the top of related product pages, and with some machine learning we could optimise which project to show against which product, and so on. It could be awesome!
Rowan:	And before we jump straight to awesome?
Alex:	We should test the idea first. We'll knock up a single prototype project page to go with some of our power tools, making it so that you can click through to all the relevant product pages. We can then grow it from there.
Rowan:	Right – power tools makes sense given that we're keeping a close eye on that area at the moment. And our hypothesis?
Alex:	A couple of those I guess: people will land on the project page from Google, and some of those people will go on to make relevant purchases. We shouldn't expect too many people to land before we've done the SEO but it's not like this is an expensive experiment.
Rowan:	Sounds like a plan, or at least the start of one. I'd like to get some research and design on the case, and I'll add it to the experiments board. And you could ask for some feedback in tomorrow's planning meeting if you think you're ready.

Implicit in this conversation is a shared understanding of Lean Startup's *build-*

measure-learn loop:

1. **Build** something that will allow a hypothesis to be tested quickly and cheaply
2. **Measure** the impact of that increment, usually in terms of customer behaviour – customer arrivals, clicks, and purchases in our example
3. **Learn** from the experience, refining the idea and its implementation if the data suggests that this is a worthwhile direction to take, or trying a different approach if the data is less encouraging

Typically, the "something" is a product increment or refinement. Early in a startup's life however, the search is not for improvement, but for viability. The startup is in search of its *minimum viable product* (MVP) – its test of *product/market fit* – and it is prepared if necessary to change direction (or *pivot*) rather than over-invest in an idea whose big assumptions can't be validated positively.

In short, Lean Startup describes a discipline of hypothesis-driven business experimentation. If – and it's an imperfect analogy – DevOps is Agile reimagined by engineers exposed to the Theory of Constraints, then Lean Startup is the same but by product managers and entrepreneurs with an interest in Lean. With its focus on *validated learning*, it is arguably the most right-to-left of all the frameworks covered in this chapter.

Despite its association with a very 21st century style of business – not always digital but certainly fast-moving – Lean Startup incorporates a lot of 20th century process improvement knowledge. In particular, build-measure-learn is a modern incarnation of the *plan-do-check-act* (PDCA) cycle popularised decades ago by Shewhart and Deming[57]. It should come as no surprise therefore that some of the tools of Lean Startup have found their way back into the world of organisational change.

One of those tools is the hypothesis. Here's Alex's idea framed as a hypothesis, Lean Startup-style:

> **We believe that** suitably-hyperlinked DIY project pages
> **will result in** increased product sales.
>
> **If successful, we might expect to see**:
>
> - Potential customers landing on these new pages
> - Existing users navigating to these new pages from inside the site
> - Onward navigations to related products
> - Product purchases, as a direct or indirect result of these new

pageviews

That describes pretty well the goals of both Alex's hypothetical "really going to town on it" vision and much the more immediately testable prototype the team will initially build. Indeed, the prototype might very legitimately be described as the beginnings of an MVP – it's not entirely beyond the realms of possibility that a viable business could be created out of the idea.

With some discussion, a hypothesis such as the one above might be all that's needed for the first iteration of work to be planned out. In our example, they might make this a safer experiment to test by skipping the second assumption – ie that existing users will navigate to these new pages from inside the existing site – and make no changes to existing pages. What they won't skip is the instrumentation (i.e. the measurements and dashboards that show them how the system is being used): they'll make sure that whatever users do or don't do with the test page, the team will get to know about it.

Used in conjunction with a development capability built with the kinds of patterns described previously in this chapter, Lean Startup 'wraps' the development process, keeping it well fed from the upstream side, attentive to customer need and behaviour, and continuously validated downstream. With the right kind of leadership, it can foster a sense of stewardship, where teams consciously work on the system for its impact. This helps them avoid becoming mere 'order takers', a mentality that sounds sensible enough until the team finds itself lacking the autonomy to steer its way out of a grotesque spiral of decline.

Lean Startup is often the catalyst for other tools to be brought into the product management space. One natural choice is Design Thinking; you could say that Lean Startup + Design Thinking gives you Continuous Discovery (and a lot else). Another excellent fit is Kanban, often used to manage experiments that may require unpredictable periods of time to achieve meaningful results (and so not easily coupled to Sprint rhythms).

Here's the kanban board design used in my Lean Startup-inspired simulation game Changeban (Figure 17):

Agree Urgency	In progress ( )			Complete ☺
	Negotiate Change ( )	Validate Adoption ( )	Verify Performance ( )	Accepted
				Rejected
Is it **valuable**? Will we get to it soon?	Is it **feasible**? Acceptable to all parties?	Is it **usable**? Will it stick?	Is it *actually* **valuable**? Is it meeting needs?	

Figure 17. The Changeban board

Apart from the rather game-specific subdivision of the rightmost column (it's designed this way to help us keep score), this is a stereotypical board for experiment management. It has a nice correspondence with the classic product management Venn diagram, *Valuable, feasible, usable* (Figure 18):

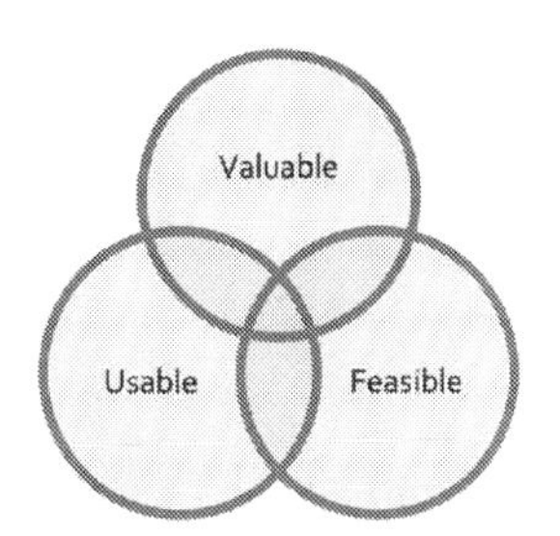

Figure 18. Valuable, Feasible, Usable

For an idea to pass all the way from 'Agree urgency' to 'Accepted' it must be:

1. Believed to be **valuable** enough to try, more so than all the other options on the table
2. Both technically **feasible** and organisationally acceptable
3. Found to be **usable**, encouraging the kinds of user behaviours we want
4. And again **valuable**, confirmed as delivering the kind of benefits hoped for

An idea needs to fail only one of these tests for it to move swiftly to the 'Rejected' part of the board.

Notice how this kind of language works equally for product ideas and for process improvements, organisational changes, and the like. We exploit this duality in the game, so that people familiar with the product space will feel comfortable applying these ideas in the change space, and vice versa. We finish with retrospective experiment design (thought-provoking in the game, a potential review technique in in real life):

> We believed <*hypothesis*>
> but found while <*activity*>
> that <*insight*>
> and rejected this idea.
>
> Had we tried <*x*>,
> we might have discovered this
> <*sooner, more cheaply, &/or more safely*>

Hindsight is a wonderful thing (and especially when it's not for real), but how many failures could have generated their learnings sooner, more cheaply, and/or more safely with better experiment design? What if we always reviewed our failures with that goal in mind? A regular opportunity to do just that is described in chapter 5.

Right to left: A grand unification theory for Lean-Agile

This chapter contained a lot of detail, but in this summary let's see how our continuous, outcome-oriented, pull-based, and right-to-left perspective on the frameworks introduced in this chapter really does help explain how they support each other:

1. **Scrum**, continuously iterating on and self-organising around goals (short term outcomes) in the pursuit of longer-term outcomes – product vision, the team's mission, broader organisational objectives, and so on

2. **Kanban**, making progress on outcomes visible, concentrating effort on the outcomes that matter most, fostering a focus on completion, and making the pursuit of fitness for purpose more tangible

3. **XP** and **DevOps**, right across development and production, providing the infrastructure of process, practice, and technology necessary to accelerate feedback on the delivery of outcomes

4. **User Story Mapping, Jobs to be Done**, and **BDD**, keeping users, their authentic needs, and their desired outcomes in full view, prioritised and developed on a continuous, just-in-time basis

5. **Service Design Thinking**, continuously discovering needs and identifying outcomes, creating the conditions in which imaginative solutions can be conceived
6. **Theory of Constraints**, continuously, strategically, and proactively identifying and addressing the bottlenecks that limit the effectiveness of any delivery process, even when built from well-tested patterns
7. **Lean Startup**, pursuing business viability through deliberate and continuous experimentation, managing for impact (outcomes again), finding and continuously refining a business model that enables customer outcomes to be sustained

Our right-to-left perspective helps also to explain the leadership roles involved in digital delivery and the collaborations between them. For example:

- Product leadership takes responsibility for the 'why' of products and services, understood in terms of customer needs and outcomes and developed just in time (broad themes for the longer term, more clarity on near-term opportunities)
- Technology leadership takes responsibility for the 'how', delivering and supporting solutions that demonstrably meet needs and enable outcomes
- The separation of these responsibilities makes the 'what' – i.e. what gets built – the fruit of collaboration. Options are generated from all sides, the innovations coming in the interplay between diverse ideas, convergence happening just in time, in the light of both recent learning and future direction.

In the next chapter, the frameworks will be larger and consequently not so easily described in terms of their individual contributions and collective synergies. Nevertheless, a right-to-left perspective is not only possible, but powerful.

Reflections

1. How do you catalyse self-organisation and collaboration around goals? How do you cause that process to be repeated reliably in the pursuit of longer-term objectives?
2. How do you prevent left-to-right tendencies (expectations of linear and implementation-driven processes, with commitments made prematurely) from dominating in contexts where a more right-to-left approach (outcome-oriented, iterative, and just-in-time) would be

more appropriate?

3. How do you ensure that teams maintain a clear sense of purpose?
4. How do you keep teams manageably small and still with the range of skills and capabilities necessary for self-sufficiency?
5. How do team-level governance mechanisms engage with those of the wider organisation?
6. Across your product development value stream, by what explicit means do you maintain attention on flow? Is work pulled into activities that have capacity available, or pushed downstream when an activity step is completed? Through what coordination mechanisms does that happen?
7. How do people, process, and technology interact to create a high-feedback environment?
8. How do you discover, identify, explore, capture, organise, and prioritise *"authentic situations of need"* and their respective outcomes? How does this understanding unfold over time? When you're starting from scratch, how do you prime the pump?
9. How do you recognise and deal with the bottlenecks and other constraints that limit the overall effectiveness of your product development process?
10. How does your organisation pursue product/market fit? By what mechanisms does it encourage experimentation and learning?

---

[36] See *Inspired: How to Create Tech Products Customers Love*, Marty Cagan (John Wiley & Sons, 2nd edition, 2018). Cagan emphasises the use of outcomes in roadmaps; also he is forthright in his description of the product manager as holding a leadership role that has far greater scope than that defined for the Product Owner role in Scrum. This is not to denigrate the PO role, but rather to explain that the framework-centric role taught "by the book" does not do justice to a demanding leadership position. I would add that 'serving the process' is a trap that all leaders must take care to avoid.

[37] Daniel Mezick described Kanban as "explicit attention to flow" in *The Culture Game: Tools for the Agile Manager* (2012, FreeStanding Press). You could say that your system gets what it pays explicit attention to.

[38] On the grounds that the kanban system is there to help improve the process, it is considered something of an antipattern to name columns after roles or

organisational units, which are generally harder to change than work item states or activities.

[39] If you have not experienced this relationship between WIP and collaboration yourself, check out our simulation games Featureban and Changeban, at agendashift.com/**featureban** and agendashift.com/**changeban** respectively.

[40] *Kanban: Successful evolutionary change for your technology business*, David J. Anderson (2011, Blue Hole Press)

[41] *Kanban from the Inside*, Mike Burrows (2014, Blue Hole Press)

[42] Unfortunately, the promotion of Kanban as a tactic to deal with work that didn't suit Scrum led to the myth that Kanban was unsuited to development work. Worse, when Kanban is used only to manage the work that doesn't sit well with Scrum, it's a short step to reinforcing the wrong kind of team boundaries too.

[43] Another myth: that deployment should happen once and only once per Sprint

[44] *Epics* in Agile are just large 'stories' – named after *user stories*, covered later in the chapter – chunks of work bigger than the work items that typically flow through the system, suggesting that they will be broken down into smaller chunks.

[45] *Flaccid Scrum*, Martin Fowler (2009), martinfowler.com/bliki/FlaccidScrum.html

[46] en.wikipedia.org/wiki/Software_craftsmanship

[47] *User Story Mapping: Discover the Whole Story, Build the Right Product*, Jeff Patton and Peter Economy (2014, O'Reilly Media)

[48] *Know Your Customers' "Jobs to Be Done"*, Clayton M. Christensen, Taddy Hall, Karen Dillon, and David S. Duncan (Harvard Business Review, September 2016 issue), hbr.org/2016/09/know-your-customers-jobs-to-be-done

[49] With an expanded remit, the UK's Skills Funding Agency is now the Education and Skills Funding Agency (www.gov.uk/government/organisations/education-and-skills-funding-agency)

[50] See *Beyond Software Architecture: Creating and Sustaining Winning Solutions*, Luke Hohmann (2003, Addison Wesley Professional)

[51] *This is Service Design Thinking: Basics-Tools-Cases*, Marc Stickdorn and Jakob Schneider (BIS Publishers, 2010)

[52] The UK's Government Digital Service (GDS) has *"Start with user needs"* as design principle #1 – see www.gov.uk/guidance/government-design-principles. For emphasis, this is sometimes expanded to *"Start with needs – user needs not government needs"*.

[53] *The Goal*, Eliyahu M. Goldratt and Jeff Cox (Routledge, 3rd edition 2004)

[54] *The Phoenix Project*, Gene Kim, Kevin Behr, and George Spafford (IT Revolution

Press, 3rd edition 2018)

[55] *The Bottleneck Rules: How to Get More Done (When Working Harder isn't Working)*, Clarke Ching (Independently published, 2018)

[56] Experience teaches that "once you have it, you'll wish you had it from the beginning" applies to almost any XP or DevOps capability.

[57] See en.wikipedia.org/wiki/PDCA. See also my article *On not teaching PDCA,* blog.agendashift.com/2016/03/01/on-not-teaching-pdca/.

Chapter 4. Viable scaling

Springboard DIY has just learned that a consortium of suppliers has banded together, insisting that retailers integrate with a new shared system, and quickly. Not that they can exactly insist, but it's an offer that few retailers can afford to refuse. Rowan, our product manager, is in the office of Nicky, Springboard's Chief Information Officer (CIO), where a meeting convened to discuss the issue is breaking up.

Nicky: Rowan, you look worried!

Rowan: No, not worried, more frustrated. There's so much that we wanted to do, and now this!

Nicky: But...

Rowan: But I completely get that we should do it. Or rather, the cost of not putting it ahead of everything else is a price we shouldn't pay.

Nicky: Exactly – I couldn't have put it better myself.

Rowan: And it's not like 5 years ago, when projects like this nearly killed us. Getting a team together would involve endless interdepartmental battles, we'd take forever working out what to do, and don't get me started on how hard it was back then to get stuff out to production when finally we got round to building something. And if it involved anything external... well it makes me shudder just to think about it. Seriously, I wonder how we survived.

Nicky: And now you're part of an amazing team, and no-one doubts that you can deliver. You're going to find that you have dependencies on other product-line teams but you know that you can count on their total support. In fact, I'm sure they're all grateful that yours is the one taking the lead on this. And if there's anyone that can find some hidden advantage here that our competitors won't spot, it's you.

Rowan: Yes, they are amazing, and I'm not worried about the other teams. Let's do this!

For better or for worse, no book on the Lean-Agile landscape would be complete without some mention of *scaling* and the scaling frameworks. Scaling is about 'bigger', whether it's teams or departments bigger than a single Scrum team, or bigger systems, bigger projects, bigger time horizons, and so on.

I should mention right away that within the practitioner community, the question of whether 'bigger' and Agile should be expected to coexist happily is controversial enough; whether the answer should come in the form of a predefined framework is more controversial still.

I won't be coming down on one side or the other of that debate, but that doesn't mean that I'm dodging the issue. Rather, there are bigger things at stake! But let's not get ahead of ourselves. First, we'll look at those frameworks, starting with the slightly misleadingly-named Spotify model – a mainly structural model of organisation design – followed by a representative sample of process frameworks, majoring on the best-known of those, SAFe.

With the frameworks introduced, we'll return to an issue raised in our comparison of the left-to-right and right-to-left flavours of Scrum: the question of how adoption should work, whether of frameworks specifically or of Lean, Agile, or Lean-Agile more generally. Appropriate to this chapter's scaling theme, we'll consider the implications not just for the teams affected, but for the wider organisation too.

The Spotify model: Squads, tribes, chapters, and guilds

The Spotify model is often presented as an organisational structure. However, it is better understood not as a solution but as a case study on how Spotify responded to a unique situation, a period of growth so spectacular that it risked failure if it paid insufficient attention to how it sustained and amplified key aspects of its culture.

The model is described in a 2012 article[58] by Henrik Kniberg and Anders Ivarsson. Like all models, it's a simplification of a messier reality and difficult to reproduce faithfully. Mention it, and you'll often hear "Not even Spotify really uses the Spotify model", sometimes from people who work at the company! Nevertheless, it's an interesting example and an important model, successful in introducing some terminology that is now quite widely understood, helpfully making some cultural ideas easier to reference. It also provides certain large consultancies with something to recommend to some very un-Spotify-like clients (forgive me if I sound too cynical on that last point).

Let's build the model bottom-up, explaining Spotify's intentions as we go. Its most granular organisational unit is the *squad*, an autonomous, long-lived, cross-functional team, with its own dedicated *product owner* .

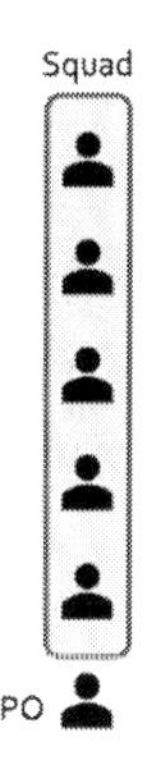

Figure 19. The squad

Squads are roughly Scrum-sized, but squads are free to evolve their own process, whether based on Scrum, Kanban, a combination of the two, or something else entirely.

Because they're long-lived, squads aren't project teams in the traditional sense. Rather, they're formed around a long-term mission, with goals to pursue, and the time to develop strong working relationships and excellent capabilities. These include the massively scalable technology that serves millions of Spotify subscribers and the ability to sustain rapid rates of change across their technology estate.

Through this model, Spotify is trying to instil a sense of the Lean Startup in its squads. To amplify that entrepreneurial spirit, these 'startups' are 'incubated' within cohesive *tribes* of up to 100 people, all working on related things, and in close physical proximity. It should be noted that Spotify have found that colocation is essential to tribal health; even they have found the model difficult to implement when it is not supported by the right kind of physical environment.

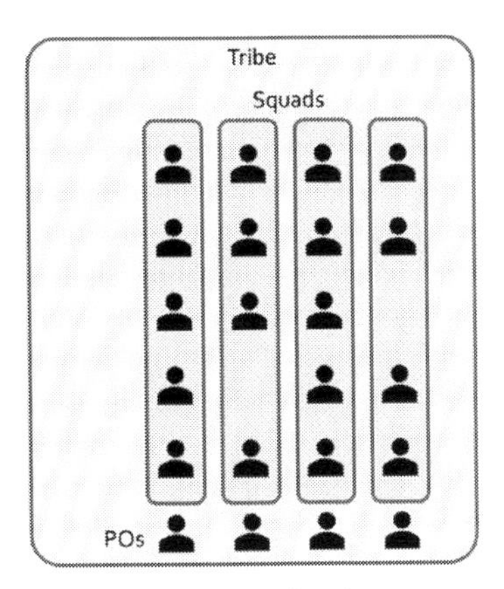

Figure 20. The tribe

Between squads – whether within a single tribe, among *alliances* of related tribes, or across tribes more broadly – communication is mostly self-organised, happening spontaneously when needed. Only for unusually large or complex projects is it necessary to convene regular and formal meetings to manage dependencies. These too can be self-organised; self-organisation and autonomy are both highly valued and relied upon. It's not a free-for-all however; product owners collaborate on the roadmaps[59] to which the squads' backlogs align.

You might be wondering how hundreds of developers can work in this self-organising way without tripping each other up. Part of the answer of course lies in DevOps – it would certainly be impossible without efficient and reliable delivery pipelines that give instant feedback whenever code changes fail to integrate well. Another vital element lies in a system architecture built out of small, independent components that can be deployed and upgraded independently.

> **Aside** (technical): This isn't a completely new idea. In the mid 90's I worked in investment banks, and in the highly competitive environment at the time, the leading banks had largely componentised and distributed their front-office trading systems, their middle-office systems were quickly catching up, and they were beginning to make inroads into their back-office systems. From front to back, these components communicated via self-describing messages that were transported asynchronously over 'middleware'. This architecture allowed each system to evolve at a rate most appropriate to its respective users (with development timeframes measured in hours to months depending on the nature of the business activity), allowed systems to be scaled globally (real-time global risk management was a key competitive advantage at

the time), and brought significant benefits in terms of system performance, resilience, and availability.

To illustrate the impact that this kind of architecture can have on organisational culture, I remember as a front office developer taking a leading role on a multi-month, multi-bank project that made significant changes to the way that trades worth billions of pounds were managed. At our bank, multiple development teams were involved 'front-to-back'. The few project managers we had were there not to coordinate us or to keep us on our toes, but to manage relationships with counterparts at other banks and with the Bank of England; the rest we did ourselves. It was a powerful lesson in what a little self-organisation can achieve in a high trust environment!

Fast forward a quarter of a century, the technology has moved on but the same principles apply. For 'distributed components' you're more likely to hear 'cloud-based services and serverless functions'. Structured communication between components is now ubiquitous, supported not by expensive, proprietary technologies but by open protocols and formats (the apps on your phone or in your browser use these all the time without you noticing). The middleware that allows services to communicate reliably and maintain data consistency even in the presence of failure is typically based on open source software and/or provided as a managed service by the cloud platform.

Whilst it's true that scalable, flexible, and resilient architectures are no longer the preserve of organisations with deep pockets, architecture remains an important discipline. Rates of system evolution, and levels of performance, resilience, scalability, security, and adaptability all depend on how components are organised and how they communicate (not unlike the human organisation!). Technology choices are perhaps less critical than once they were, but how they're used still matters greatly.

To recap, we have *squads* – small, autonomous, long-lived teams, each with their own mission – that belong to *tribes*, the 'startup incubator' of those squads. Where their interests align closely enough to justify special levels of inter-tribe cooperation, tribes may join forces to form *alliances*.

The next element in the Spotify model is the *chapter*, which gathers together people with similar skills from across the squads of a single tribe. A *chapter lead* is the line manager for other chapter members; chapter leads in turn report to the *tribe lead*.

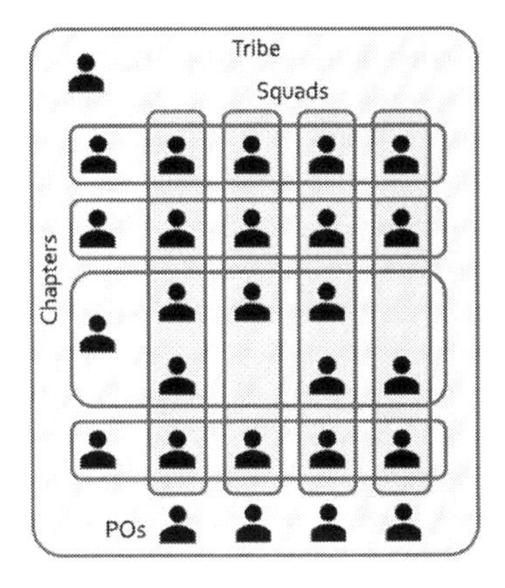

Figure 21. Chapters

Of course, chapters are not just (or even mainly) about reporting structures. The goal of the chapters is to provide coherence, knowledge sharing, problem sharing, common tooling, and so on – the kinds of things that help to reinforce and develop the tribe's individual culture.

Scaling that idea up a bit, *guilds* are the Spotify model's communities of interest. They span multiple tribes, connecting people that share some common interest, reinforcing and influencing Spotify's wider culture. Guild membership correlates somewhat with that of chapters, but people are free to join any guild. *Guild coordinators* are responsible for making sure that their guild serves the needs of its members and the wider organisation in accordance with its remit.

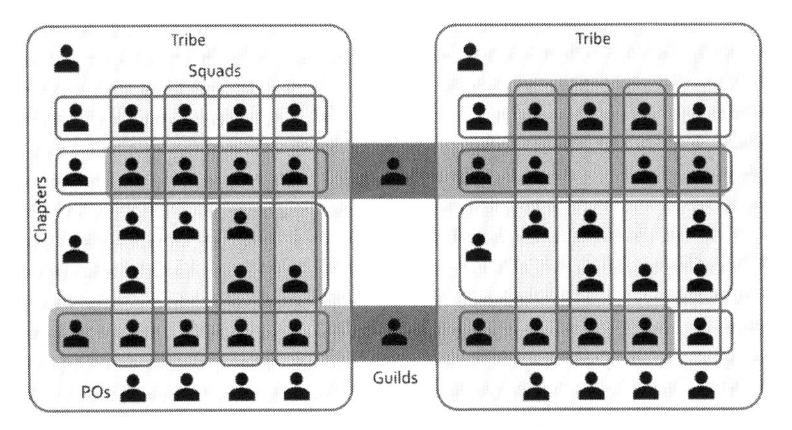

Figure 22. Guilds

In further support of Spotify's commitment to engineering excellence and process improvement there are three roles not shown on the diagram:

- *System owners* – responsible for the architectural integrity of each of Spotify's main systems (squads can make changes to multiple systems)
- A *chief architect* – coordinates architectural work across multiple systems
- *Agile coaches* – aligned to squads

In summary, and without the jargon:

- A large number of small, diverse, entrepreneurial teams, each with their own long-term mission and goals to pursue, and their own team-level ways of working
- Low-overhead mechanisms designed to maintain coherence across several cross-cutting dimensions, most notably particular culture, architecture, and product
- Founded on and amplifying values of autonomy, self-organisation, engineering excellence, and continuous improvement

Arguably, this is the true Spotify model, the one that outlasts the structural model that even Spotify have struggled to implement consistently. If you're thinking of borrowing from Spotify, wouldn't this be the better starting point?

Scaled Agile process frameworks

The term 'scaled Agile process frameworks' describes the several branded frameworks that have surfaced in the past decade or so, designed to help a team-level process scale up in one or more dimensions. To take a small but representative sample:

- Nexus™[60], which features a single Product Backlog, a single Product Owner, multiple Scrum Teams, and an Integration Team. It is a model of *horizontal scaling*, a way for a Scrum-based delivery organisation to maintain consistent relationships externally even as capacity is added or removed internally.
- Scrum@Scale®[61], which scales 'upwards' through the mechanism of the *Scrum of Scrums*, to create a hierarchy of Scrum Teams.
- Scaled Agile Framework® (SAFe®)[62], which emphasises periodic programme-level planning, a hierarchical work breakdown structure,

and coordinated teams working in a team-of-teams called an *Agile Release Train* (ART).

A comparison matrix for a much longer list of frameworks – nearly all of them Scrum-based – is maintained at agilescaling.org. I have the space here to focus on just one, and I choose SAFe, not because I am endorsing a favourite, but for these three reasons:

1. It is well known and well supported
2. As it has evolved, it has become less insistent on the use of Scrum at team level (and I needn't dwell therefore on whether or why Scrum might need these frameworks, and why there are so many of them)
3. Of all the scaled Agile process frameworks I have investigated, it is the one that most explicitly invites a right-to-left interpretation

Essential SAFe

Glossing over a certain amount of detail, I'll describe the core process of SAFe in two ways, based respectively on the left-to-right and right-to-left descriptions I used for Scrum in the previous chapter. First, the left-to-right version (or how not to describe SAFe if you can help it):

- *n* levels of backlog, where the parameter *n* is proportional to the height of your SAFe poster[63].
- Planning events of various kinds take work from one level of backlog and into the next level down, with regular opportunities for reflection and improvement. At the lowest level, teams each have their own Team Backlog, which they refresh for each timeboxed iteration.
- Groups of development teams share an integrated delivery pipeline.

Previous chapters have identified several shortcomings and risks associated with strongly backlog-driven processes. As things scale up, the additional challenge is to manage the gap between the discovery and specification work done on those backlogs upstream and the eventual customer impact downstream; the bigger the plans and the greater the number of levels, the harder that becomes (and that's assuming that this challenge is even recognised).

At the time of writing, SAFe goes further than its key competitors in explicitly supporting an alternative, right-to-left interpretation[64]. My description below is based on the pattern of *iterated self-organisation around goals* that we used for

Scrum in the previous chapter:

- Teams move toward their objectives goal by goal.
- Teams collaborate around their goals in cycles of predetermined length, at the end of which they reflect on how well their goals were achieved and look for ways to improve. They then prepare to organise around new shared and team-specific goals, the opportunity to try new ways of working, within each team and across teams.
- Each team's best understanding of the work required to achieve their current goals is represented by its Team Backlog; options for future cycles are maintained in higher level backlogs and brought forward as capacity allows.

The 'cycles' in this process are often referred to as Sprints, although it is no longer mandated that teams will use Scrum; officially the term is 'iterations'.

What makes this rather generic-sounding process definitively SAFe are these two elements:

1. **The Agile Release Train (ART)**, a team of teams (typically comprising 50+ people) that share an overall vision, a delivery pipeline, and synchronised planning rhythms, all facilitated by a Release Train Engineer (RTE). Larger SAFe implementations may have multiple release trains and some higher-level coordination between them.
2. **Program Increment (PI) Planning**, a facilitated, all-hands planning workshop that takes place at fixed intervals (typically 8-12 weeks apart), marking the boundaries of a multi-Sprint timebox.

Subject to the presence of those mandated elements, *Essential SAFe* is a minimal SAFe with a very Scrum-like 2-level backlog. Larger editions of SAFe add 'Large Solution' and 'Portfolio' levels (although it is a little misleading to say "add" here; it would be more accurate to say that the smaller editions came into existence as tailorings of the full model).

Why the controversy? Most objections fall into one of five categories:

1. Concerns over specific practices, for example that not every organisation will be brave enough or have pockets deep enough to sustain regular PI Planning
2. Concerns over the complexity of the larger SAFe editions. The SAFe 4.5 Reference Guide[65] runs to 816 pages, and even the distilled version is 416 pages long[66]
3. Concerns over the growth of a service industry built to support this

complexity; at the time of writing there are for example no fewer than 10 certifications offered by Scaled Agile Inc, with a global network of trainers qualified in delivering certified training, and above them a process for training the trainers

4. Concerns over applicability – how sure can anyone be that implementing SAFe will actually make things better?
5. Concerns over how SAFe is sold and implemented, with some evidence of bad behaviour

I'll address my remaining comments mainly at objections 4 and 5; dealing with these questions of applicability and implementation deals also with concerns over practices and complexity, objections 1 and 2. As for objection 3, commercial success represents some positive validation, but if its business model of role-based certification allows SAFe to dominate the market in the way that Scrum dominated team-centric Agile, innovation inevitably suffers.

Some of the controversy is both answered and stoked by the nine SAFe principles:

1. Take an economic view
2. Apply Systems Thinking[67]
3. Assume variability; preserve options
4. Build incrementally with fast, integrated learning cycles
5. Base milestones on objective evaluation of working systems
6. Visualise and limit WIP, reduce batch sizes, and manage queue lengths
7. Apply cadence, synchronise with cross-domain planning
8. Unlock intrinsic motivation of knowledge workers
9. Decentralise decision-making

That's a pretty good list; with no more than minor reservations, I agree with them all. In fact, it is fully acknowledged that many of these principles owe a debt to Don Reinertsen's *Principles of Product Development Flow*[68], a book widely regarded in the Lean-Agile community as foundational. But, given that there are many ways in which these principles can be applied, the question of whether SAFe is a good choice can only be answered in context.

SAFe is likely to be a poor choice wherever either of these conditions apply:

1. These principles are already in evidence to some reasonable extent, and SAFe's design choices are likely to worsen rather than improve

the organisation's current situation in these respects were they to be introduced, or

2. There are simpler, less disruptive, and more direct ways to achieve whichever of these principles have the most leverage in the current situation

Those criteria alone should cause many potential adopters to think twice. Part of the controversy therefore stems from a fear – justified or otherwise – that SAFe will be sold into situations that won't benefit, the cure if not worse than the disease then at least inferior to some of its available alternatives. However, that still leaves a sizeable addressable market of enterprise clients happy to pay for an enterprise solution to their legacy problems. The size of that market is enough to explain the existence of SAFe, not to mention its several imitators and competitors.

Engagement models and the challenges of change

If it seems that I have been equivocating, taking care not to come down on one side or the other, you'd be right. But frankly, the choice of framework – if any – matters very much less than the approach the organisation takes to adoption. Just as we saw in the case of delivery, there are left-to-right and right-to-left approaches to change.

First, the traditional, 20th century approach of *managed change*, a left-to-right, solution-driven, and implementation-focussed approach:

- A solution is chosen – SAFe, in this case
- Authentically or otherwise, a case is made for the implementation's urgency
- The solution is rolled out, and in the face of *resistance to change*, the plan followed
- And then the aftermath: disappointment with the results, staff disengagement, and the realisation that the world has moved on meanwhile and the process must be repeated

A caricature no doubt, but still horrifying for how easily recognisable it is. So then, what might a right-to-left, needs-based, and outcome-oriented approach to change look like? Here is one possible answer, a SAFe-flavoured Agendashift[69]:

- Establish a shared sense of direction, exploring purpose, objectives, and key challenges

- Identify needs – one starting point for which might be an examination of the organisation in the light of the nine SAFe principles – and the obstacles that prevent those needs from being met
- Agree on outcomes – not just arbitrary goals, but the kind of outcomes that might be achieved when the most important of those obstacles are removed, overcome, or bypassed
- On a just-in-time basis, prioritise outcomes and generate a range of options to realise them, using SAFe as a source of ideas – not to exclude other sources, but valuing the coherence presumably gained by referencing a framework
- In manageably small chunks of change and through a combination of direct action and experimentation (choosing between those approaches according to the level of uncertainty and risk involved), begin to treat change as real work: tracking it, validating its impact, and reflecting on it just as we would for product work

What I have sketched out are two highly contrasting kinds of *engagement model*, models for how change agents do their work. To be effective in the organisational change space, an engagement model must do three things:

1. Help to structure the work of change agents – facilitators, consultants, coaches, or employees whose remit includes the encouragement of change
2. Help the client organisation engage its staff meaningfully in change-related work, inviting high levels of participation
3. Help those parts of the client organisation that are undergoing deliberate change to engage constructively with the rest of the organisation, so that all sides will thrive

The best engagement models address all three concerns in ways designed to encourage staff engagement. The worst have the opposite effect – disengagement – and that's a real problem for Agile. For SAFe specifically, much of the negative attention directed on it is for these reasons:

1. The engagement models of certain big consultancies are focussed mainly on executives, to whom they pitch a Spotify + SAFe solution by default, seemingly with little regard to applicability or alternatives. Some of these consultancies leave the implementation work to other firms, so they have remarkably little skin in the game themselves.
2. When these consultancies do offer to lead this work themselves, left-to-right rollout models are still the norm. Moreover, these models

are referenced in SAFe training aimed at leaders – not exactly mandated but still the default.

3. The likelihood of leaving the transformed delivery organisation plugged into a legacy project/programme management organisation that has no reason not to continue thinking in left-to-right terms. With any right-to-left aspirations quickly falling by the wayside, the net effect is that not much fundamentally changes, projects and programmes continuing much as they did before.

It doesn't have to be this way, which is why I don't dismiss the branded process frameworks out of hand. If my refusal to see them as designs to be implemented makes me a subversive, well I can live with that.

More positively though, we're now seeing a flowering of what have been dubbed *Open Social Technologies*, a much broader, more diverse, and still complementary array of frameworks that address a range of organisational concerns in ways that the process frameworks on their own simply cannot. They're open in multiple ways:

1. Not only are they well documented, they share substantial parts through open source, Creative Commons, and similar mechanisms (including release into the public domain), to the extent that a suitably-experienced practitioner could with effort reproduce and even improve on it without necessarily licencing whatever conveniences might be available to them on a commercial basis.

2. They're open not just to extension (a natural property of any framework) but also to modification and replacement, something that many branded frameworks actively discourage. To be truly open, there must be no mandated practices; instead an attitude of *"core or better"*[70] prevails, enabling both local adaptation and community-driven innovation.

3. They are non-exclusive; like the patterns of chapter 3, they combine in multiple, interesting, and exciting ways, the composition often greater than the sum of the parts.

4. They are highly responsive to context, capable of being transposed from one social or organisational context to another, producing perhaps radically different outputs according to the situation. To achieve this, they're *generative*, such that outputs are generated, organised, prioritised, and developed by participants – predetermined neither by the framework nor the facilitator.

As social tools, they help people to work together, empowering them with decision-making authority, building social capital up, down, and across the organisation and beyond its four walls. This describes both how they work when they're being used deliberately and the kind of organisational legacy they tend to leave behind. For the engagement models in particular, this internal consistency is an explicit design goal, one that contrasts sharply with the dissonance and disengagement too easily triggered by traditional approaches to change.

A small selection of relevant frameworks that demonstrate these properties:

- The engagement models Agendashift and OpenSpace Agility™ (OSA)[71]
- Clean Language (chapter 5), which via Agendashift or on its own is valued by parts of the Lean-Agile community as a coaching protocol. Its heritage however is in psychotherapy – a powerful demonstration of the responsiveness to context described in point 3 above!
- Liberating Structures (chapter 6), a library of facilitation patterns, also referenced by Agendashift
- The large-scale collaboration framework Open Space Technology (OST, chapter 6)

Organisations have needs too

To finish this chapter, we return to the third of the three required features of effective engagement models, which is to:

3. Help those parts of the client organisation undergoing deliberate change to engage constructively with the rest of the organisation, so that all sides will thrive

Decades of study have gone into what it takes for organisations and all its parts thereof to thrive. Key elements include:

- A sense of **identity** and **purpose**, discovered, maintained, and reaffirmed
- **Strategy** processes that set future direction and objectives, and **policy**-setting processes that affirm organisational constraints and preferences in accordance with identity, purpose, and strategy
- Various **management** and **monitoring** processes that keep organisational and **delivery** processes on track
- **Audit** and **intelligence** processes that keep all of the above informed about what is happening inside and outside the organisation

This structure comes from the field of *management cybernetics*, which is founded on the *viable system model* (VSM[72]) developed by Stafford Beer in the 1960s and 1970s. Beer sought a universal model for viability, and motivated by a desire to avoid using a terminology that might mislead, he referred to the processes above not by familiar names but as numbered 'systems' (System 1, System 2, etc). Admirable as his intentions might have been, it does have the unfortunate effect of making this important model a little opaque to outsiders. Therefore, if you would like to dive deeper into this body of knowledge I would steer you in the direction of something accessible, and I can wholeheartedly recommend Patrick Hoverstadt's book *The Fractal Organization*[73].

A key feature of VSM is its recursive or fractal nature. A viable organisation must have these elements not just at the top level, but at every level of organisation, and they must operate coherently between levels and across them. This in turn has crucial implications both for the design of feedback mechanisms (a topic for the next chapter) and for structure.

For the organisation wanting to make rapid progress (the fledgling digital organisation being a prime example), structuring around product lines means that customers and staff can readily identify each other, their needs, and their purpose. Where sheer scale dictates, other structural dimensions such as customer segment, geography, or technology may come into play above or below the product dimension, but care must be taken to minimise the number of organisational boundaries spanned in any value stream.

The traditional structure based on functional silos often offers the worst of all worlds. Functional structures tend to:

- Diffuse product and service responsibilities across multiple functions
- Maximise rather than minimise friction from a customer, product, or organisational alignment perspective
- Reduce resilience, because from the comfort of their own internally-consistent models, silos are often slow to recognise and respond to the sheer diversity of external needs and challenges[74]

These are real problems even in times of relative stability; when organisations established along silo lines attempt anything new or try to increase the pace of development, they quickly find that organisational boundaries impede flow not just across delivery processes, but in strategy, issue resolution, and innovation processes too[75]. In the presence of more nimble competition and perhaps unseen disruptive forces, a structure that might once have been seen as a strength may turn out to be a potentially fatal flaw.

That said, a restructuring – however necessary – is not a panacea. Matrix organisations, for example, are often criticised for their complexity, for their tendencies to induce conflict, and for leaving people feeling pulled uncomfortably in multiple directions. However, as a safe-to-fail experiment and a temporary halfway house in a transition from a functional organisation to something more product-aligned, then why not? Bottom line, no organisation is ideal forever and you keep working both with and on the one you have. More concretely, you start with the people you have and the diverse mix of competencies, capabilities, and organisational experience that they bring – individually and collectively – and take it from there.

Attention to coherence across boundaries is the difference between an organisation that makes sense on paper and one that really thrives. It's an active, two-way thing: teams need both to align to higher level objectives and to be given the opportunity to participate in their creation. In times of rapid change this is even more important; when Agile adoptions look only inward and fail to build trust externally, they are often surprised to find not only that their existing problems haven't gone away, but that new and potentially serious issues have materialised as interfaces, relationships, and identities change.

Blaming the organisation and its managers for starting from the wrong place is of course not the answer. A team-centric or software-centric Agile can be forgiven for not having much to offer the wider organisation (let alone an adequate theory for it), but as a digital leader you don't have that luxury. As you grapple with difficult issues, a little cross-border empathy can go a long way; it will help you in both your search for understanding and your quest for mutually beneficial solutions.[76, 77]

To be concrete and practical about this, the next chapter introduces some deliberately outward-looking and boundary-spanning tools, two of them in the shape of feedback loops that you can implement to augment, re-energise, or replace some of your strategy and monitoring processes. Both of them are both framework-agnostic and multi-level; you can use them inside your chosen framework, across the boundary of your Agile adoption (if such a boundary exists), or outside it.

Reflections

1. How do you maintain the architectural integrity of your key systems in the presence of rapid change? Similarly, how do you maintain an engaging and appropriately consistent experience for the users of your products?
2. How do you cultivate a sense of entrepreneurialism in your teams?

3. How do you maintain a level of cultural coherence across your organisation, consistent with the autonomy of each organisational unit?
4. How do you sustain *"iterated self-organisation around goals"* beyond single teams? At a technical level, how is the work of multiple teams brought together? How is the work of multiple teams aligned to shared product and business objectives? How do they collaborate?
5. What brings structure to the work of those whose remit is to encourage change in your organisation?
6. By what means are employees encouraged to participate meaningfully in change-related work?
7. When parts of the organisation are undergoing significant change, what keeps them constructively engaged with the rest of the organisation?
8. What are your organisational structures and how do they help customers and staff to identify each other, their needs, and purposes? In what ways do your organisational boundaries encourage and hinder flow?

---

[58] *Scaling Agile @ Spotify with Tribes, Squads, Chapters & Guilds*, Henrik Kniberg and Anders Ivarsson (2012), blog.crisp.se/wp-content/uploads/2012/11/SpotifyScaling.pdf

[59] Spotify's roadmaps are of course outcome-based. Embracing the level of uncertainty involved, outcomes may be framed as *bets*. The bigger the upside and the less dependent they are on untested assumptions, the better the bet.

[60] The Nexus™ Guide, www.scrum.org/resources/nexus-guide

[61] Scrum@Scale®, www.scrumatscale.com/

[62] Scaled Agile Framework® (SAFe®), www.scaledagile.com/

[63] I jest of course. SAFe's much commented upon posters are downloadable from scaledagileframework.com/posters/.

[64] To be clear, I'm not suggesting that other frameworks aren't capable of being given a right-left-interpretation, just that they don't make it as explicit. Change here would of course be welcome.

[65] *SAFe 4.5 Reference Guide: Scaled Agile Framework for Lean Enterprises*, Dean Leffingwell (Addison Wesley, 2018)

[66] *SAFe 4.5 Distilled: Applying the Scaled Agile Framework for Lean Software and Systems Engineering*, Richard Knaster and Dean Leffingwell (Addison Wesley, 2018)

[67] *Thinking in Systems: A Primer*, Donella Meadows (White River Junction, 2008)

[68] *The Principles of Product Development Flow: Second Generation Lean Product Development*, Donald G. Reinertsen (2009, Celeritas Pub)

[69] *Agendashift: Outcome-oriented change and continuous transformation*, Mike Burrows (New Generation Publishing, 2018); agendashift.com

[70] *"Core or better"* is a nod to one of the 'commitments' of the Core Protocols (Jim McCarthy and Michelle McCarthy, www.mccarthyshow.com/online): "*I will use the Core Protocols (or better) when applicable".*

[71] See the website openspaceagility.com and the *The OpenSpace Agility Handbook*, Daniel Mezick, Mark Sheffield, Deborah Pontes, Harold Shinsato, Louise Kold-Taylor, and Mark Sheffield (Freestanding Press, edition 2.2, 2015)

[72] This VSM, the Viable System Model, is not to be confused with the VSM that we saw in chapter 1, Value Stream Mapping.

[73] *The Fractal Organization: Creating Sustainable Organizations with the Viable System Model*, Patrick Hoverstadt (John Wiley & Sons, 2008)

[74] A motivation for my mention of diversity is a powerful concept from cyberneticist Ross Ashby, namely *requisite variety*. Informally (without the maths): *In order to deal properly with the diversity of problems the world throws at you, you need to have a repertoire of responses which is (at least) as nuanced as the problems you face.* Source: *What is requisite variety*, Dan Lockton, requisitevariety.co.uk/what-is-requisite-variety/

[75] See the book that with the benefit of hindsight proved not only to be well ahead of its time but a catalyst for new thinking on innovation: *The Innovator's Dilemma: When New Technologies Cause Great Firms to Fail*, Clayton M. Christensen (Harvard Business School Press, 1997)

[76] Small but quite disproportionately vocal parts of the Agile community have a rather awkward relationship with anything management-related. Unhelpfully, blaming managers and their organisation for starting from the wrong place (and worse, assuming bad motives) seems to be a default position there.

[77] In their book *Conscious Capitalism* (for which a full reference is given in the next chapter), John Mackey and Raj Sisoda note the relationship between emotional intelligence, systems intelligence, and spiritual intelligence. Without doubt, empathy and Systems Thinking (en.wikipedia.org/wiki/Systems_theory) can be mutually supportive in a powerful way; either one practised deliberately can lead to the other. See also the *learned optimism* of *positive psychology* (en.wikipedia.org/wiki/Positive_psychology).

Chapter 5. Outside in

It's a big day for Chandra, tech lead for our team at Springboard DIY. It's her turn at the company's quarterly product strategy briefing, a meeting she has attended before but never as a speaker. Her task is to give a strategic overview of her team's product area.

Normally this would be a job for Rowan, the team's product manager. Chandra is however undaunted; she and Rowan are close colleagues, and they work hard together to ensure that everyone in the team has the opportunity to contribute to a strategy that both encompasses and integrates both of their respective domains. The speaking part isn't a worry either, since most product briefings follow a well-practised pattern that she is happy to follow:

1. **Customer**: An overview of that product area's key customer segments, some interesting insights into the needs of customers in those segments, any trends that they're noticing or trying to influence, and so on
2. **Organisation**: An affirmation of the relationship between the customer overview and Springboard's mission and strategy, explaining the impact (intended or otherwise) of recent, current, and planned corporate initiatives
3. **Product**: Recent, current, and planned developments in the product space, including changes to the product range, new features, marketing campaigns, and so on, also financial performance
4. **Platform**: Recent, current, and planned developments in technology, intellectual property, and operational capability
5. **Team(s)**: The opportunity to celebrate what's happening internally in terms of team achievements (eg key deliveries), people (eg key hires, notable individual accomplishments), and process performance

Chandra isn't forced to follow this 'outside-in' structure, but she knows that it works. If she can make her briefing sound both coherent and compelling, it will be a boost to everyone's confidence. And she's got this!

What's happening?

After four chapters majoring on right-to-left thinking, this penultimate chapter introduces the concept of the outside-in review (in fact we'll cover outside-in reviews of several kinds). What's happening? Why the need for a new metaphor?

The preceding chapters have been about bringing a right-to-left, needs-based, and outcome-oriented perspective to the delivery process – a natural perspective for Lean, a healthy perspective for Agile, and a proven model for digital. In these final two chapters we step back from the delivery process and introduce two additional and complementary perspectives that are less about process and more about the organisation. Both of them remain anchored in the constant themes of this book, needs and outcomes.

For this "outside in" chapter, instead of working our way upstream from those key moments of value creation, we start from outside the organisation, establishing a holistic view of needs, challenges, and desired outcomes. Then we drill through its various layers, working our way in towards the heart of its operations. Taking this route, any organisational misalignments should reveal themselves.

We will cover:

1. The outside-in Service Delivery Review (OI-SDR)
2. The outside-in Strategy Review (OI-SR)
3. NOBL's Organisational Charter
4. Wardley mapping

I have selected these tools in particular (and in the case of the first two, developed them) not because they are the best known but because they most clearly demonstrate the outside-in perspective. Better known to the Lean-Agile community are the Business Model Canvas[78] and Lean Canvas[79]; perhaps in the light of this chapter you may wish to investigate or re-evaluate these too.

The outside-in Service Delivery Review (OI-SDR)

In one form or another and over several years, the Service Delivery Review is a meeting I've both attended and implemented; it's well established in traditional IT operations[80], and – as championed mainly by the Kanban community – gaining recognition in development circles too. In systems terms, it supports monitoring and control, and with the right participation, management and intelligence too.

Here's how the outside-in agenda works in this setting:

1. **Customer**: What we're hearing from our customers, via the customer helpdesk, user research, user-submitted feedback and so on; fresh insights arrived at since the previous meeting; customer-related assumptions validated, invalidated, in the process of being tested, or soon to be tested; customer satisfaction
2. **Organisation**: Things happening outside our team that we need to know about and potentially participate in; progress on organisational blockers taken away from past meetings; organisational performance
3. **Product**: Recent, current, and planned experiments in the product space – how they're performing (if live) or their hoped-for impact; product performance overall
4. **Platform**: Recent, current, and planned developments in technology, intellectual property, and operational capability; incidents; planned maintenance; technical debt reduction activities; platform performance and capacity
5. **Team(s)**: A closer look at the delivery pipeline, staffing, and the performance of the delivery process

This structure may seem a little unconventional, but let me advocate for it under the six headings of context, alignment, participation, data, learning, and completion:

Reason 1: Context

Because it starts outside the boundary of the organisational unit in question, everything identified as happening inside the boundary is properly contextualised. The risk of *suboptimisation* (the technical term for local optimisation at the expense of the whole) is greatly reduced.

Reason 2: Alignment

Between 'layers' – between customer and organisation, or between product and platform for example – any contradiction, confusion, misalignment, or tension will be laid bare for all to see. That can be uncomfortable for those concerned, but the resolution process can be a source of real progress. The act of making tough decisions out in the open can be highly instructive; if policy is clarified, future decisions will be easier to make and perhaps even delegated.

Reason 3: Participation

Each agenda item is led by the person or people best equipped to represent current activity and knowledge in that area. For example:

- **Customer**: Product manager or product owner, user researcher, UX designer
- **Organisation**: A senior manager, and perhaps other stakeholders external to the team. In UK Government digital services, the Service Manager (a hybrid business/technology role) is a great fit here.
- **Product**: Product manager or product owner
- **Platform**: Tech lead, tech support lead, test lead, delivery manager, representatives of dependent services
- **Team**: Delivery manager, team leads, coaches, perhaps HR if appropriate to the review's scope

The OI-SDR doesn't need to be a big meeting; my first one involved just six people huddled around a small table in an informal meeting area. Notice however the potential for close interaction between people who represent not just multiple disciplines but multiple organisational levels, including some that you might not expect to see in (say) a team retrospective or leadership team meeting. I have come to appreciate this aspect so much that I now look for it and plan for it – more on this in chapter 6.

Reason 4: Data

Representatives bring data in the form of individual facts (discoveries, recent events, and so on) and aggregate metrics. Instead of chasing a limited number of 'killer metrics' that offer little real insight on their own, the meeting's structure invites a broad range:

- **Customer**: Net Promoter Score (NPS) or similar; helpdesk calls and hours spent on them; customer complaints, endorsements, and reviews; user growth and retention
- **Organisation**: Financial metrics, progress against relevant organisational objectives, and so on
- **Product**: Usage analytics; funnel metrics; market comparisons
- **Platform**: System performance and capacity metrics (along with plans to keep capacity ahead of anticipated demand – another good reason for the outside-in review); new capabilities and capabilities under development
- **Team**: Lead time distribution, throughput, and work in progress (in any combination – remember Little's law!); quality metrics (defects escaped to production, for example); data on blockers and their

impact; staffing levels; skill distribution and development

For some participants, the outside-in Service Delivery Review may be their first experience of gathering, interpreting, and presenting metrics. They might find it challenging at first, but it's a useful skill to learn, and then there's the satisfaction of seeing the team's actions play out in metrics over the course of subsequent meetings. And that's a key point: metrics here are not the basis of systems for reward or supervision (metrics-driven versions of which lead almost inevitably to dysfunction[81]), but are collected and shared by self-organising teams in support of their own work.

A good set of metrics will do several things:

- Open a window into operations and give early warning of potential problems
- Give actionable insights (metrics that never prompt action are unlikely to be worth the cost of preparing them)
- Show how system behaviour is changing over time – in response both to interventions and to factors beyond the team's immediate control
- Warn when any metric is being chased to the unacceptable detriment of other things (for example lead time at the expense of throughput, throughput at the expense of quality, new signups at the expense of retention, and so on)
- Speak to the values and actual experience of customers, the team, and other stakeholders

The meeting's facilitator is responsible for ensuring that each agenda item will be appropriately supported by data, and that the overall set meets everyone's needs. This may create valuable coaching opportunities when the OI-SDR is first instituted and as new participants come on board.

Reason 5: Learning

The outside-in Service Delivery Review provides an ideal opportunity to track progress on experiments. Insights derived from short-running experiments can be shared after completion; long-running experiments can be tracked from meeting to meeting. Reviewing experiments as part of each agenda item has the effect of normalising experimentation, improvement, and innovation, making them natural aspects of doing business. Sustaining change like this through system-reinforced expectations is a piece of deliberate organisational design, and it's one that sends a powerful message about the organisation's values. And remember: there is nothing at all inevitable about cycles of change – you must design for them!

It's very easy to share a post-experiment insight and still miss much of the potential learning opportunity. Our training game Changeban (mentioned in chapter 3) finishes with an exercise in *double-loop learning*[82]:

> For each activity in the process, describe an experiment that rejected a product or improvement idea:
>
> > We believed <*hypothesis*>
> > but found while <*activity*>
> > that <*insight*>
> > and rejected this idea.
> >
> > Had we tried <*x*>,
> > we might have discovered this
> > <*sooner, more cheaply, &/or more safely*>.

Completing this two-part template even for an imaginary experiment teaches that there are multiple levels of learning on offer, some focussed on the experiment's chosen subject, others on the processes by which insight is achieved. Doing it for real helps to reinforce the idea that each failed experiment or rejected idea represents a positive decision if managed effectively, a concept that risk-averse organisations often find difficult to embrace.

Reason 6: Completion

In my description of the outside-in Service Delivery Review you may have noticed a recurring pattern: "Recent, current, and planned". That's right: the OI-SDR isn't just outside-in, it's also right-to-left! Instead of reviewing actions or experiments in subsequent meetings in the order that they were first raised (the classic recipe for interminably long and ineffective meetings), we focus first on recently completed work (with needs met and learning captured), then on work that we can get over the line. Priorities for upcoming work comes last, and always with a careful eye on capacity.

In choosing which individual issues or pieces of work to discuss, some discipline is required on the part the meeting's facilitator (and/or secretary) and its participants. It is not the purpose of this meeting to review every piece of work, but if something is brought to a meeting, it is reasonable to expect that it will be carried over to subsequent meetings unless specifically agreed otherwise. As a guide, any issue or work item expected to span multiple meetings should be brought to it; these are the items most at risk of losing their way. With ground rules like these agreed, you'll soon settle into an effective pattern that covers the important items quickly and sees them through to a definitive conclusion.

How often you hold these meetings will depend on context. My own experience suggests that weekly is usually too often, and quarterly not often enough. To begin with, try monthly or four-weekly. Use any pressure to hold them more frequently – to align with two-week sprints for example – as an incentive to automate your metrics!

The outside-in Strategy Review (OI-SR)

The idea of approaching strategy from the outside in is not new. Here, for example, is *Forbes* staff writer Wendy Tanaka in her 2010 article *The Value Of An 'Outside-In' Strategy*[83]:

> *The premise is that consistently successful companies start with an external market orientation and vigilantly study customer trends in order to design their strategy. This is called "outside-in" thinking, using customer trends as a guide post for product and service development.*
>
> *Conversely, an "inside-out" strategy is one that relies upon an internal orientation. It starts by asking what a company can do with existing resources, and looks to streamline operations through right sizing and repressed spending. While this approach can create short-term shareholder gains, an internal focus limits a company's ability to notice and adapt to market changes.*

More bluntly: whilst it's important continually to be building on internal capabilities, a strategy that's oblivious to what's actually happening outside risks ruin.

Our kind of outside-in Strategy Review (or OI-SR) has all the advantages of the outside-in Service Delivery Review – context, alignment, and participation most especially – but the two reviews should be seen not as alternatives but as complementary. As we shall see, it quite literally asks a different set of questions. As a workshop, it has this 5-session structure:

1. Agree on a timeframe
2. The outside-in questions
3. From Obstacles to Outcomes (FOTO)
4. Structure a plan
5. Developing options

Session 1. Agree on a timeframe

This one's easy – just ask! For us to achieve something worthwhile, how long do we need?

As it happens, this question is also asked by Celebration-5W[84], an exercise that I use frequently as the Agendashift workshop's kickoff exercise. In a nutshell: working in table groups, participants describe the Who, What, When, Where, and Why (the journalistic 5W's) of the post-strategy celebration:

- **Who** are we, and who else will be celebrating with us
- **What** are we celebrating – the accomplishments that our strategy will bring about
- **When** are we celebrating – how long we will need to achieve something truly celebration-worthy
- **Where** are we celebrating – a venue that says something about who we are and what we've achieved
- **Why** are we celebrating – the essence of what makes this whole effort worthwhile

Clearly, the question of *"How long do we need?"* is answered by the When part; the other parts helping to validate the answer.

This exercise doesn't necessarily need a fancy template – simple headings down the page will do – but this one (Figure 23) designed by my friend Mike Haber works very nicely. To fill it in, it helps to start with the When (bottom right) and work anticlockwise; readers will naturally start top left with the Who or in the middle with the Why.

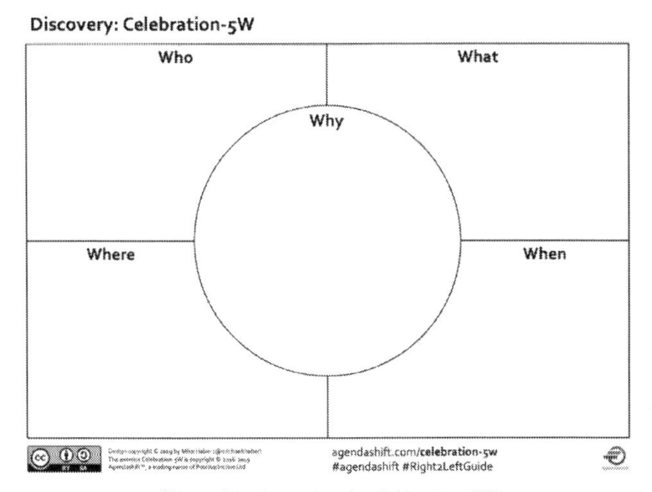

Figure 23. A template for Celebration-5W

Sometimes the sponsor will give me a planning horizon in advance as part of my initial brief. Usually we agree that it would be interesting to hear what others think, and we kick off with the Celebration-5W exercise anyway. I don't require table groups to synchronise their answers after sharing their work in the post-exercise debrief; working to different timelines adds a little extra diversity, and that's no bad thing.

Session 2. The outside-in questions

This session asks a series of questions, worked on in table groups one question at a time, following the same outside-in sequence as the Service Delivery Review. The '*When*' in these questions is given by the timeframe agreed in session 1, so the answers represent goals to pursue over that period:

1. **Customer**: What's happening when we're reaching the right customers, meeting their strategic needs?
2. **Organisation**: When we're meeting those strategic needs, what kind of organisation are we?
3. **Product**: Through what products and services are we meeting those strategic needs?
4. **Platform**: When we're that kind of organisation, meeting those strategic needs, delivering those products and services, what are the defining/critical capabilities that make it all possible?
5. **Team(s)**: When we're achieving all of the above, what kind(s) of team(s) are we?

You may recognise the style of these questions. They're not exactly the canonical *Clean Language* questions – more on these when we get to the next session – but there's certainly something "cleanish" about them. As facilitators, we discipline ourselves to ask open and non-leading questions, minimising the risk that the conversation will end up being driven not by the participants' collective intelligence but by the facilitator's assumptions. The way in which the later questions contain echoes of previous questions is also a little cleanish; here it's done to help keep everything aligned. Our hope is that we're giving participants the best possible opportunity to build and explore their own coherent models of the topic or challenge in question.

The first question – for the customer layer – deserves some special attention as it sets up all the rest. Also, it's a little more subtle than the others as it begs two further questions and suggests a third:

1. Who are those *right customers*?
2. What are their *strategic needs*?
3. Why us (the 'we' of the question) and not some other alternative, a competitor, say?

After considering those, focus on the *"What's happening"* part.

If you're starting with a blank page, you may want to allocate some time to identifying the right customers and then to their '*strategic needs*' – the needs that best define your mission. For subsequent reviews, you might agree to narrow the focus to particular kind of customer or a more specific set of needs. For example, if you plan to expand into a new market segment, customers there may have a new set of needs for you to explore.

Then, in this future that you're building, why is it *your* organisation that is meeting those right customers' strategic needs and not some alternative, a competitor or new entrant, say? It's easy to give glib answers to a question like that, but an important goal of strategy is to identify the position you want to occupy with respect to other participants, actual and potential. That's hard, because nothing stays still!

With this simple two-part reflection you can build some awareness of movement:

1. How has your business environment changed in the past couple of years, say? What has changed for all the main actors involved (customers, suppliers, channels, competitors, peers, regulators, and so on) in that time?
2. Now project that forward: outside of what you can control, what changes might you need to prepare for? Given those possibilities,

how do you best position yourselves for success? And when the currently unknowable eventually reveals itself, what makes you think that you will be ready to respond?

With that in mind, revisit and revise your previous work. I don't pretend that this is the last word on manoeuvre-based strategy, but it's a start! If you're interested in exploring this concept further, I refer you to *Patterns of Strategy,* co-authored by my friend and collaborator, Patrick Hoverstadt[85].

The remaining questions can simply be taken at face value, but for a more rigorous examination, you can try taking an outside-in approach even to the individual layers, reflecting on their respective external relationships first. For example, if your scope is the whole organisation, 'outside' means the outside world, suggesting questions such as these:

- What kind of organisation are we in our relationships with community, society, and the environment? (We'll see these important themes mentioned again in chapter 6 in relation to Servant Leadership; see also *Conscious Capitalism*[86])
- What kind of organisation are we in our relationships with the business, economic, and regulatory ecosystems in which we operate?
- Taking these together with our previous answers, what kind of organisation are we?

Session 3. From Obstacles to Outcomes (FOTO)

You've painted a broad-brush picture of where you'd like to get to; now you must return to your present reality and think about how you move forward from here. This is classic coaching territory – consider for example John Whitmore's classic coaching model, GROW[87]:

- **G**oal: What you would like to achieve, where you would like to go, how you would like things to be
- **R**eality: How things really are
- **O**ptions: How might you make progress from here?
- **W**ill (or **W**ay forward): What do you commit to doing?

Translating this kind of structure from personal coaching into workshop facilitation, our go-to tool is the Clean Language-inspired coaching game 15-minute FOTO[88], FOTO being an acronym for 'From Obstacles to Outcomes'. Very briefly:

1. Table groups prepare and prioritise a list of obstacles that lie in the

way of their goals – in this case obstacles to their strategic goals as identified in answer to the questions of the previous session. Groups are asked to ensure that every participant can speak to at least one obstacle.

2. Participants take turns for a few minutes at a time in the roles of client, coach, scribe, and observer. The client chooses an obstacle from the prepared list, and using Clean questions from the cue card (Figure 24), the coach guides a conversation that explores the client's mental landscape of obstacles and outcomes. The coach is careful to reflect the client's own words, and the scribe writes down *"anything that sounds like it might be a desired outcome"*, the bar kept deliberately low so that nothing is excluded prematurely. The observer acts as safety officer and ensures that the conversation stays on track, focused mainly on outcomes rather than digging unproductively into obstacles.

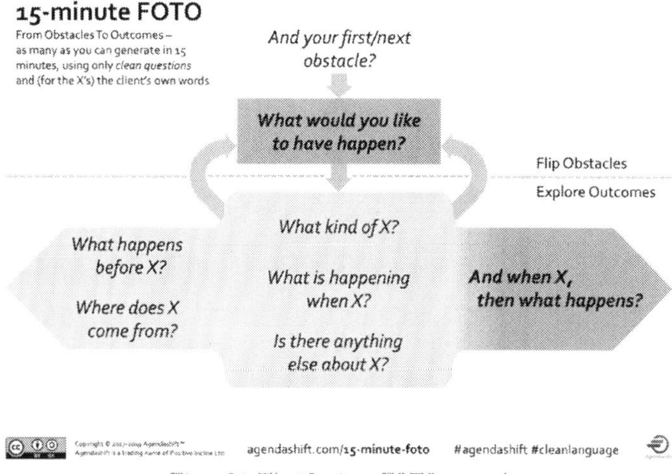

Figure 24. The 15-minute FOTO cue card

Whatever your favoured facilitation approach, I strongly urge that you focus on generating a broad portfolio of outcomes rather than picking favoured solutions at this stage. As alluded to in the previous chapter, no-one likes to have solutions imposed on them, and it's important not ruin the opportunity for options to be generated just in time by the people most affected by any change and best equipped to maximise its impact. What better way to mobilise than to invite interested people to self-organise around an outcome with a license to innovate? There will plenty of opportunity for that outside of this particular meeting.

Session 4. Structure a plan

Session 3 generates lots of output; the purpose of session 4 is to organise it into something that other people will understand and respond to. Formats will vary according to the amount of detail generated but here's a selection of the tools I use most often:

- *Affinity mapping* and *dot voting*: Put all the individual pieces of output onto sticky notes, group similar items together, and label the main groups. Using coloured dot stickers or marker pens, participants are given a small number of votes each (3 is typical), thereby prioritising themes.

- *Plan on a page*: A simple 3-column plan on an A3-sized sheet of paper can be surprisingly effective. I like to put short term outcomes (urgent items and quick wins) on the left, long term outcomes (long term objectives, aspirations, and values) on the right, and between them, your medium term outcomes (intermediate objectives, signs that you're winning).

These techniques are not mutually exclusive. You could for example employ a 2-dimensional structure, with emergent themes or the five layers of the outside-in agenda labelling rows down the page and the three time horizons labelling columns across the top. Our online resources include a template you can download[89] (Figure 25).

Outside-in Strategy Review (OI-SR)

			Outcomes		
	Objectives	Obstacles	Short term	Medium term	Long term
Customer					
Organisation					
Product					
Platform					
Team(s)					

agendashift.com/oi-sr-template
#Right2LeftGuide

Figure 25. An OI-SR template

Session 5. Developing options

This last session is an opportunity to practice some steps that will be repeated time and time again as the strategy is executed (or rather, *deployed* – *strategy deployment* being a topic for the final chapter). I summarise here a four-step process described in much more detail in chapter 4 of *Agendashift*, the Elaboration chapter:

Step 1. Choose an outcome to focus on

Naturally, we take a just-in-time approach and choose an outcome (or one outcome per table group) that we want to pursue soon; the rest are better left until their time comes. Mainly for the sake of illustration, we further ask participants to limit themselves to outcomes that are open to a range of possible actions; we know that a hypothesis-based approach is likely to be highly appropriate in such cases[90].

Step 2: Generate some options

On their own (silently), each participant generates two or three options, trying to make them as diverse as possible. The table groups then create a shortlist – five or so – of the more promising options, again trying to keep the list diverse.

Step 3: Choose one

Which option is most likely to significantly outperform, relative to the others and relative to the investment required? *"What would have to be true for this option to be fantastic?"*[91]

Step 4: Frame a hypothesis

With this template, the option is framed as a hypothesis:

> **We believe that** <*actionable change*>
> **will result in** <*meaningful outcome*>.
>
> **If successful, we might expect to see:**
>
> - <*observable impact*>
> - …
> - <*observable impact*>

When introducing the technique to groups that haven't tried it before, I ask for at least three kinds of observable impact, evidence that the idea is working as expected in the direction of the chosen outcome.

Hints:

- What initial signs might you see?
- What would you need to see before you would say that it is

successful?

Step 5: Design an experiment

Finally, the hypothesis is developed into an experiment, capturing:

- **Risks** (to the downside): possible undesirable impact
- **Potential benefits** (or *upside risks*): Things we're not specifically planning for but will encourage if we begin to see them
- **Assumptions**: Things we don't yet know
- **Dependencies**: Things we haven't yet put in place
- **Pilot experiments**: Actions that we can take that are much faster, cheaper, and safer than our overall experiment; each pilot experiment mitigates a downside risk, amplifies a potential benefit, tests an assumption, or resolves a dependency – ideally delivering some business benefit along the way
- **People**: everyone potentially impacted by our idea, especially people who will be asked to work differently, modify policies, or reconsider their priorities in some way

A popular Lean technique is to capture an experiment's design on an *A3*, a single sheet of A3-sized paper. There are as many A3 templates as there are Lean consultants, and here's mine, with spaces for all the above elements:

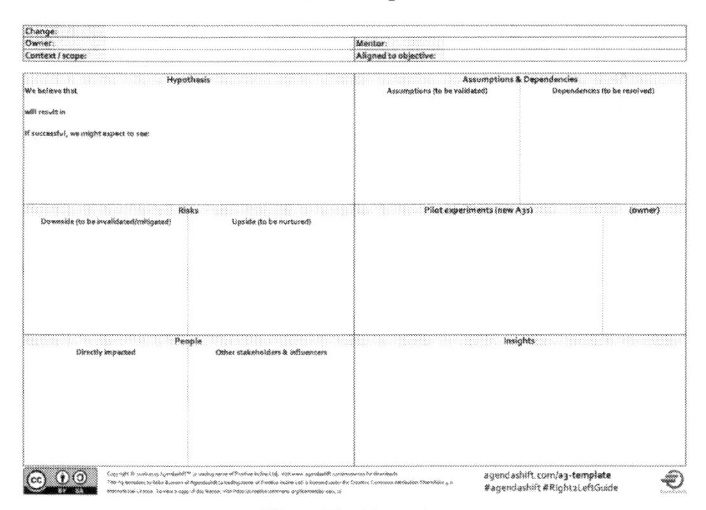

Figure 26. A3 template

This can be downloaded from our resource pages[92].

After you've filled in your A3, it's good to perform a quick cross-check that helps to identify any gaps. Review your experiment's different impacts, imagine the insights you'd like your experiment to generate (there's a space for those too, though you can't complete it yet), and then use your judgement to verify that your pilot experiments are capable of delivering them.

Once you have a decent first draft, your next step is to show it to a representative sample of the people that you identified. They will expect you to present your ideas clearly; in return they will help to test your thinking. This is what A3 is all about.

The wholehearted organisation

The two outside-in reviews I've described have quite different roles: one helps to sustain progress on the strategy while the other deliberately opens it up to challenge, and with it perhaps the status quo. One way in which that kind of challenge is important is articulated beautifully by the renowned architect Christopher Alexander:

> *"A thing is whole according to how free it is of inner contradictions. When it is at war with itself, and gives rise to forces which act to tear it down, it is unwhole. The more free it is of its inner contradictions, the more whole and healthy and wholehearted it becomes."*[93]

I love the idea that as leaders and trusted advisers we can choose to be in the business of helping organisations to be more *wholehearted* – less at war with themselves, their contradictions identified and owned so that they can be resolved in some satisfying way. By way of analogy, if we improve our delivery processes by removing impediments to flow, then we improve our organisations by removing impediments to alignment. To that very worthwhile end, the outside-in reviews are specifically designed to bring misalignments, impediments, and contradictions to the surface.

We should not however take for granted that this alignment is well directed. Fundamentally, the viability of any organisational unit – at every level from team to whole organisation – depends on the effectiveness of its processes for (i) deciding what's important, and (ii) ensuring that the unit's work remains focussed on the right things. If yours aren't good enough (and that's more common than you might think), try our outside-in forms of the strategy and service delivery reviews – first as a check or a refreshing change of perspective, and then to enhance or replace your existing governance meetings.

We start from the outside to create opportunities for identity, purpose, mission, and strategic goals to be rediscovered and reaffirmed. Sometimes the process can seem painful, but if your organisation faces a crisis of identity, purpose, or mission, or if its strategic goals are unclear, it is well worth the discomfort. More typically though, it's a source of encouragement and renewal: confirmation that broadly we're on the right track but still able to incorporate new learning, acknowledge gaps in knowledge and capability, and pursue fresh opportunities.

NOBL's Organisational Charter

To finish this chapter, brief introductions to two more tools that could well be described as outside-in. The first of these comes from NOBL (pronounced "no-bell"), a consultancy specialising in organisation design, which for them means the application of Design Thinking to organisations. The second is Wardley Mapping, a highly visual approach to strategy developed by Simon Wardley.

NOBL's Organisational Charter has the following high level structure:

1. **Purpose**: the reason why we choose to work together indefinitely.
2. **Strategies**: the bets we're currently making to fulfil our Purpose.
3. **Structures**: the division of work and resources we need to execute our Strategies.
4. **Systems**: the tools we need to align behaviour across our Structures.

Very much like our outside-in Strategy Review, it is clear that each element of this structure speaks to its predecessors, which means that it's crucial to start in the right place. Here we start with *purpose*, which NOBL likes to uncover with these two questions:

1. *What do we want to change about the world and why?*
2. *How can we use our collective skills to make change and what will the world look like when we succeed?*

For further detail on each of the four sections, a downloadable template, and guidance on its use, go to *How We Describe an Organization*, academy.nobl.io/how-to-define-and-describe-an-organization/.

Wardley Mapping

Wardley Mapping[94] is just one tool from a suite of strategy tools developed and open sourced by Simon Wardley. The idea is beautifully simple, the first two steps of which have a lot in common with the value stream mapping approach we took in chapter 1:

1. Start with a user need
2. Working backwards, draw a *value chain*, the chain (or network) of components required to meet that need

The value chain differs from the value stream map in two important respects:

1. In contrast with the activity-based value stream we saw in chapter 1, this value chain identifies the capabilities, resources, and infrastructure needed
2. Instead of building it right to left with a strong sense of left-to-right flow, we build it top down. It is anchored at the top with a user need; underneath is drawn a network of dependent components, items positioned vertically according to how visible they are to the user – the less visible they are, the further down the page they go.

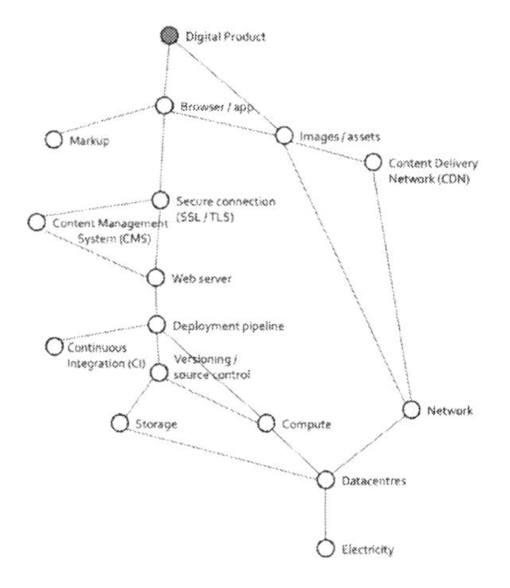

Figure 27. A value chain for a generic digital product[95]

Next, maintaining their vertical positions, components are repositioned horizontally against a maturity scale comprising these four product life cycle stages:

1. **Genesis** – the initial idea
2. **Custom built** – a one-off realisation of the idea, likely one of many possible solutions
3. **Product** (and Rental) – components can be bought, sold, rented in, or rented out; different implementations are differentiated to a significant degree
4. **Commodity** (and Utility) – implementations are very easily sourced

and highly interchangeable

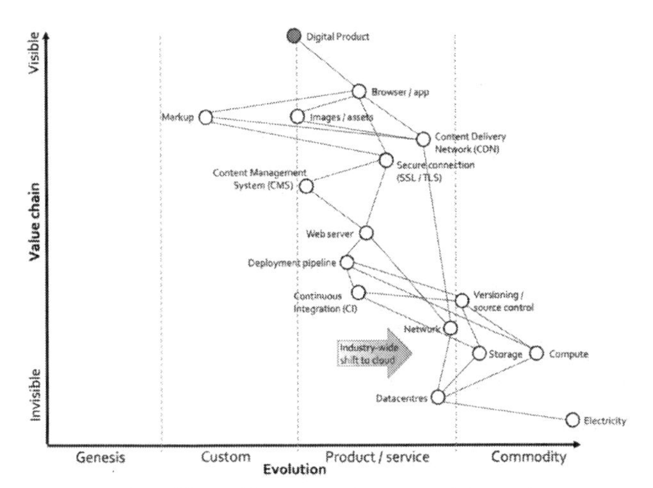

Figure 28. An Wardley map for a generic digital product

At Springboard DIY for example, an electric drill is a product to be sold; their digital presence is a one-off, built from a range of custom, product, and commodity components.

Note the tendency for components to progress rightwards along the product maturity scale over time. For example, a few years ago, a database technology would be chosen after a careful selection process, and then hosted on the organisation's own hardware. Today however, database technologies are highly interchangeable, and increasingly they are consumed as cloud services. In effect, the database product has become commoditised and is on its way to becoming a utility (a state the storage hardware has already achieved). Instead of choosing a database on features, we pay a provider capable of giving us the level of service we want at a price point we're prepared to pay.

Wardley Mapping comes into its own as a tool for strategy when it is used to identify opportunities to move components proactively in order to gain competitive advantage. For example, if you can reduce your dependence on custom solutions and replace them with products or commodities, you may gain a cost advantage over your competitors (the trade-off of course being that you lose differentiation, but that may be acceptable and perhaps even an advantage). Conversely, you might productise an internal capability for sale or rent, stealing a march on your rivals, perhaps even turning them into customers. Or perhaps you see an opportunity to start from the left, disrupting the established order with something completely new.

Outside in, and out again

Just because we start from customer needs and work inwards doesn't mean that we are always following a lead given by our customers. We absorb intelligence from outside, make our strategic choices internally, and see the effects of our decisions ripple out in every direction.

The effectiveness of this outside-in process depends on:

- People from a range of disciplines – customer-facing, product, technical, and business – engaging skilfully with the outside world and maintaining awareness of operations internally
- Intelligence, both internal and external, being disseminated quickly, suffering the minimum of delay, distortion, or dilution before it can be evaluated and acted upon
- Frequent opportunities for informed decision-making, decisions being made at the right time, not too early (closing off options prematurely) or too late (when options are limited and with the risk of disruption to existing work)
- The ability to execute, not in the limited sense of following a fixed plan over a period of time, but concentrating effort and aligning forces where it most matters, remaining alive to the possibility that circumstances may change without warning

A fast delivery process not only insufficient, it's worse than useless if misdirected. As Russell Ackoff once said:

> *"The righter we do the wrong thing, the wronger we become".*

Reflections

1. How do you keep your reviews of current operations grounded in their proper customer and organisational context?
2. How do you keep your organisational unit's range of capabilities and perspectives aligned? Who participates in that process? What data do they bring?
3. What keeps your organisation reminded of the need for experimentation? How does it ensure that learning is captured and shared?
4. How do you decide what gets tracked? What gets discussed first?
5. For a future timeframe of your choosing, the outside-in questions:

1. **Customer**: *What's happening when we're reaching the right customers, meeting their strategic needs?* (And: *Who are those right customers, what are their strategic needs, and why us?*)
2. **Organisation**: *When we're meeting those strategic needs, what kind of organisation are we?*
3. **Product**: *Through what products and services are we meeting those strategic needs?*
4. **Platform**: *When we're that kind of organisation, meeting those strategic needs, delivering those products and services, what are the defining/critical capabilities that make it all possible?*
5. **Team(s)**: *When we're achieving all of the above, what kind of team(s) are we?*

6. Whether between the layers of the outside-in questions or within them, how are your organisation's internal contradictions dealt with? What contradictions do you face now?
7. What is your organisation's purpose? Explore that with NOBL's questions:
    1. *What do we want to change about the world and why?*
    2. *How can we use our collective skills to make change and what will the world look like when we succeed?*

Which of your components or capabilities might be progressed deliberately through the product maturity life cycle in order to achieve competitive advantage? What opportunities for innovation or disruption can you identify?

---

[78] *The Business Model Canvas*, www.strategyzer.com/canvas/business-model-canvas

[79] *The Lean Canvas*, leanstack.com/leancanvas

[80] The IT service management framework ITIL (www.axelos.com/best-practice-solutions/itil) calls it the Service Review.

[81] See *Measuring and Managing Performance in Organizations*, Robert D. Austin (John Wiley & Sons, 1996) – a book that is more health warning than design manual!

[82] en.wikipedia.org/wiki/Double-loop_learning

[83] *The Value Of An 'Outside-In' Strategy*, Wendy Tanaka, www.forbes.com/sites/ciocentral/2010/12/01/the-value-of-an-outside-in-strategy/

[84] Celebration-5W is licensed under the Creative Commons Attribution-ShareAlike

4.0 International License; go to agendashift.com/**celebration-5w** to obtain the materials.

[85] *Patterns of Strategy*, Patrick Hoverstadt and Lucy Loh (Routledge, 2017)

[86] *Conscious Capitalism: Liberating the Heroic Spirit of Business*, John Mackey & Raj Sisoda (Harvard Business Review Press, 2014)

[87] John Whitmore, *Coaching for Performance: GROWing Human Potential and Purpose: The Principles and Practice of Coaching and Leadership* (Nicholas Brealey Publishing, 4th ed., 2010).

[88] 15-minute FOTO: agendashift.com/**15-minute-foto** and chapter 1 of the Agendashift book

[89] The outside-in Strategy Review (OI-SDR) template is licensed under the Creative Commons Attribution-ShareAlike 4.0 International License; go to agendashift.com/**oi-sr-template** to obtain the materials.

[90] We explore the question of appropriateness in chapter 2 of *Agendashift*, drawing both inspiration and a fantastic workshop exercise from the complexity framework Cynefin, en.wikipedia.org/wiki/Cynefin_framework. Informally, if you can imagine that asking for suggestions from five experts might easily generate ten ideas, none of which delivers your outcome completely, it's not hard to see that you would be well advised to take an iterative, experiment-based approach rather than a linear approach in which a single solution is decided up front.

[91] *Playing to Win: How Strategy Really Works*, A.G. Lafley and Roger L. Martin (Harvard Business Review Press, 2013)

[92] The Agendashift A3 template is licensed under the Creative Commons Attribution-ShareAlike 4.0 International License; go to agendashift.com/**a3-template** for the materials.

[93] *The Timeless Way of Building*, Christopher Alexander (OUP USA, 1980). Although being about architecture and the built environment, this book was a key inspiration to the patterns movement in software (and from there to chapter 3 of this book) – see en.wikipedia.org/wiki/Software_design_pattern. Quotation copyright © 1979 Christopher Alexander, reproduced with permission of the Licensor through PLSclear.

[94] Source: Simon Wardley, *An introduction to Wardley (Value Chain) Mapping*, blog.gardeviance.org/2015/02/an-introduction-to-wardley-value-chain.html, published in 2015 under a Creative Commons 3.0 ShareAlike license.

[95]Figures 27 & 28 are lightly adapted from a 2019 article by Chris Adams, *Plotting a path to a greener web with Wardley mapping*, www.thegreenwebfoundation.org/news/plotting-a-path-to-a-greener-web-with-wardley-mapping/. The article reports work done at the Green Web Foundation

with the support of the German Prototype Fund and is published under a Creative Commons Attribution license.

Chapter 6. Upside down

Rowan and Chandra are back in Nicky's office, both looking a little stunned. After a long pause, Chandra is first to speak.

Chandra: Joint heads?

Nicky: Exactly!

Chandra: But that's huge!

Nick: Not really. The customer-facing teams that currently report to me will report to you jointly. All our governance processes remain, and I will urge you not to make any structural changes until we've seen how well this works. The real point is that everyone can see that not much happens in Digital without you two, and frankly, you have more to offer to your colleagues in your respective disciplines than I do now.

Rowan: What about you?

Nicky: Oh, don't worry about me! I'll still have plenty to do, not least making sure that you have everything you need in order to be successful. Honestly, it clarifies things, and every colleague that I've sounded out feels the same way. We're talking about some closer integration with the business, and with this change you'd have the opportunity to participate fully in those discussions.

Chandra: Do we need to agree right away?

Nicky: Soon would be good, but you should have a private chat about it first. In fact, it wouldn't hurt for you to come back with your ideas on your shared priorities – assuming you're both up for it of course! You're very familiar with Springboard's priorities already, but I suspect that you'll find fresh a perspective on them.

Servant Leadership

The Scrum Guide makes an indirect reference to Servant Leadership in its description of the Scrum Master as *"a servant-leader for the Scrum Team"*, and goes on to describe various mainly Scrum-specific ways in which the Scrum Master can serve the Product Owner, the Development Team, and the Organisation.

But what does Servant Leadership mean? We could start with some of the ways people respond intuitively to the phrase, for example:

- *"Serving people"*, or *"Serving the team"*
- *"Leading by example, behaving well"*[96]
- *"Unblocking all the things and getting out of the way"*

These thoughts may be admirable but as definitions of leadership they raise more questions than they answer. We turn therefore to what is widely regarded as the most authoritative source (as well as one of my all-time favourite books), Robert K. Greenleaf's *Servant Leadership: A Journey into the Nature of Legitimate Power and Greatness*[97]. As you can tell from the title, the book is not short of ambition, but nevertheless it's also a remarkably humble book, the work of a practicing Quaker who served as trustee at a number of organisations of international repute.

It starts with this premise:

- The days of the 'job for life' will soon be over; people will move from job to job, **choosing the organisations and leaders that serve them best**

That might seem almost obvious now, but in the 1970's when it was written, it was as prescient as it was insightful. Greenleaf goes on to address his prediction with a masterful piece of Systems Thinking:

1. The first responsibility of the Servant Leader is to **help others to be successful** – removing impediments, ensuring that basic needs are met
2. For people to remain engaged, the Servant Leader must **help others find autonomy and meaning in their work**, together discovering, developing, and pursuing the organisation's values, mission, and purpose in society
3. For this process of transformation to be sustained indefinitely, Servant Leaders must **help develop Servant Leadership** in others

If you're looking for a high-level description of the kind of leadership needed in organisations whose long-term viability depends on an engaged workforce, then this is it. The necessarily challenging *"values, mission, and purpose in society"* part notwithstanding, it is not just a recipe for top-down leadership; most of it is readily accessible to anyone who wishes to pursue a leadership path at any organisational level, and regardless of whether their aspirations are recognised formally in their roles.

There is significance to the sequence in which the three aspects of Servant Leadership are presented. The Servant Leader must be servant first, leader second; the leader who isn't meeting needs is a leader whose legitimacy and authority are diminishing until that day when they find themselves with no-one to lead.

In the terms of Aaron Dignan's excellent recent book, *Brave New Work*[98], Servant Leadership is both *"people positive"* and *"complexity conscious"*:

- **People positive**, because it is so focussed on other people (and so optimistically), with success measured in terms of their needs and outcomes
- **Complexity conscious**, because through its people it helps the organisation discover and become what it needs to become, not as a one-off project but as an open-ended journey

It's a worthwhile mental exercise to review the leadership roles identified in chapter 1 through a Servant Leadership lens. To help you perform that exercise for yourself, here they are again:

- **Product leadership**, conceiving and evolving products that meet the needs of willing customers
- **Technical leadership**, designing products that can be delivered and supported effectively and that are rewarding to use
- **Market leadership**, connecting people and products, managing demand
- **Process leadership**, finding better ways to operate and manage the delivery process
- **Change leadership**, catalysing change through sponsorship, experimentation, facilitation, coaching, and coordination
- **Executive leadership**, removing structural impediments, ensuring that the organisation pulls together, now and into the future

How do people in these roles help others to be successful, help others find autonomy and meaning in their work, and help develop Servant Leadership in others? And let's bring it home: how do *you* do these things? Around you, what mechanisms are in place to promote these things? And what obstacles stand in the way?

It's worthwhile also to look at your digital initiative through Servant Leadership lens. What is it about, and over what kind of timescales? A website? A one-off project? A catalyst for longer-term change as the business redefines its mission? And then: What does that mean for people? What opportunities for leadership might be thrown up? What organisational obstacles will get in the way of longer-term goals?

To that last question, I'll give you the standard warning: some of the toughest obstacles you may encounter will lie in the 20th century 'best practices' employed in many HR, finance, procurement, legal, and project management functions – practices that despite all claims to the contrary seem almost deliberately designed to perpetuate short-term and individualistic behaviour and to suppress innovation and learning.

Whilst it's true that they can present some real challenges to Servant Leadership specifically and to right-to-left approaches generally, I would counsel against defeatism or delay. More optimist than idealist, I would repeat the old saying *"Where there's a will there's a way"*, knowing that countless organisations have found their own ways to mitigate, overcome, remove, or simply ignore problems that once might have seemed insurmountable.

Having encouraged you, let me also you warn against a grave mistake: that of attempting to address your organisation's so-called cultural challenges separately from its goals, postponing the pursuit of the latter in some narrow (and likely futile) attempt to fix the former. It should be clear that baked into Servant Leadership – and into all six chapters of this book – is the understanding that they're best tackled together, and continuously.

Inverting the pyramid

Henshaw & McCallum's 1855 organisational diagram of the New York and Eyrie Railroad is an unusually beautiful artefact, certainly in comparison with the org charts we typically see today. By today's standards it is upside-down, with the Board of Directors as the *"fountain of power"* (their words) at the bottom of the picture and the organisation growing plant-like out of it.[99]

It is more than a little ironic therefore that the modern-day metaphor of the *inverted pyramid*[100] should be so powerful. This elegant idea is easily conveyed in a single sentence:

- The organisation exists to support those who serve its customers

Visually (Figure 29), the CEO is depicted at the bottom vertex of the triangle;

the flat top side is where staff engage with customers. Any structure in between is there to support the layers above, those closer to the customer.

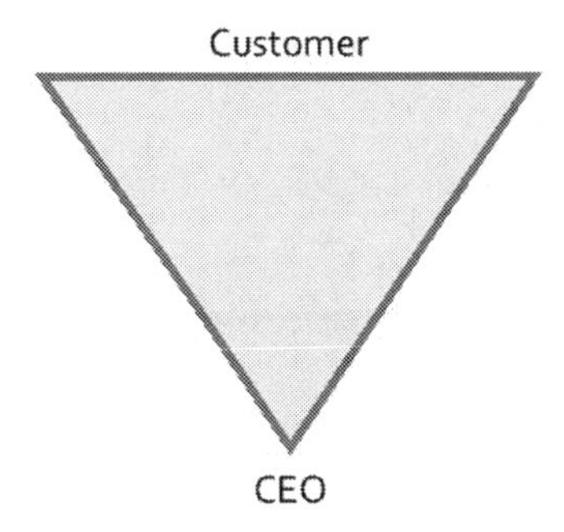

Figure 29. The inverted pyramid

Much as I love the metaphor, I'm going to play around with this picture (Figure 30). In a bid to make it work at any organisational level (or range thereof, making it *scale free*), let's replace the CEO at the bottom with a process, **Mission**, and the customer interface at the top with another, **Delivery**. Sandwiched between those two processes, I add a third, **Capability**, which has the job of ensuring that Delivery has both the individual capabilities and the level of capability required in order for it fulfil its mission. Finally, and adding some Lean-Agile flavour, we have the three organisational processes beginning with **Discovery** and ending in **Validation**. Because the intelligence that Discovery and Validation generate may inform any of the processes, I show them as shared by all three.

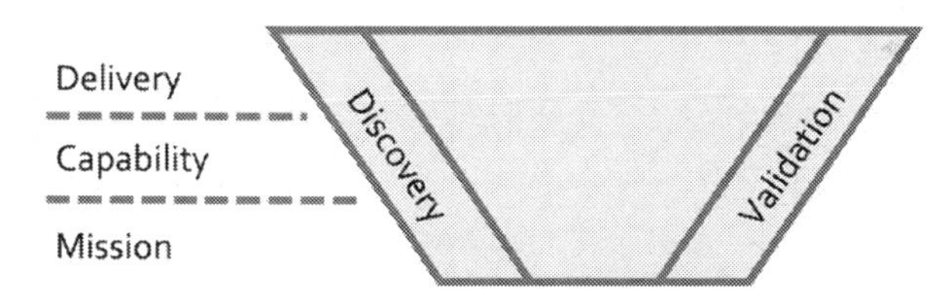

Figure 30. The inverted pyramid (modified)

In a little more detail:

- **Delivery** is about continuously meeting needs via a collaborative process of knowledge discovery, through which teams keep learning more about their customers, the relationships that their customers have with their products, and about their approach to their own work
- **Capability** supports Delivery, ensuring on behalf of all stakeholders

that delivery processes continue to perform as expected, that improvement is sustained, and that new capabilities are developed in good time

- **Mission** supports **Delivery** and **Capability** with clear objectives, an overall sense of purpose, and the values, principles, and policies that guide decision making and help the organisation maintain a coherent and engaging identity
- **Discovery** builds empathy with customers and other stakeholders and keeps the three processes fed with intelligence
- The prospect of **Validation** causes assumptions to be articulated up front, and because the possibility of failure is acknowledged, risks are addressed proactively and learning is maximised

In this simple model we have most of the ingredients for a viable organisation (chapter 4) that meets chapter 2's Lean-Agile definition: the pursuit of flow, the means to deal with complexity, the collaboration, and the continuous discovery, delivery, and learning. There are however a couple things the model does not make clear:

1. Where does adaptability come from?
2. Does collaboration apply only to delivery?

To the first question, we'll see in a moment how adaptability follows from a particular kind of leadership behaviour. This behaviour happens also to promote collaboration, but to address the second question more fully we'll return to a question hinted at in chapter 5 but left answered: who participates in each decision-making forum?

Inverting control

The search for organisational adaptability is not a modern phenomenon. No doubt you've heard the phrase *"No plan survives contact with the enemy"*. It's a pithy rendition of this longer sentence: *"No plan of operations extends with any certainty beyond the first contact with the main hostile force"*. It was written in 1871 by Helmuth von Moltke the Elder, a Field Marshall in the Prussian army.

Von Moltke was grappling with a fundamental challenge: how can you be successful as an organisation when preconceived plans are so fragile? His solution: to train leaders at every level to refrain from issuing detailed instructions that would almost certainly become meaningless very quickly. Instead, their commands were to be statements of intent and expressions of desired outcomes, to which their subordinates would then respond. Through this radically new military doctrine known today as *mission command*, von

Moltke achieved a previously inconceivable reconciliation: alignment on shared objectives with autonomy on the ground. This combination would serve people well even in the most extreme of environments.[101]

Fast forward nearly 150 years, and let's suppose now that the intent is expressed not only by leaders to subordinates, but in the opposite direction. When this is happening, the subordinate is demonstrating their grasp of the situation, showing initiative, perhaps attempting something new and innovative. The commander has a choice to make: to trust them to get on with it, to provide support, or to suggest alternative courses of action. The commander's correct response is highly situational, but in each case the two participants are mutually accountable, the one for his or her actions, the other for providing a supportive environment for personal and team development in situations where not only careers but lives may be at stake.

This is Marquet's *leader-leader* model, in which mission command meets Servant Leadership. Authority is explicitly moved to whichever parts of the organisation have the relevant information, without in any way absolving senior leaders of their ultimate responsibilities. Staff and crew meanwhile (Marquet was a submarine captain in the US Navy) are invited to vocalise their intentions, an explicit encouragement for self-management and self-organisation at individual and team levels. In time come habits of *working out loud*, the almost continuous stream of *"I intend to..."* statements no longer needing the permission of the question *"What do you intend to do?"*. Self-organisation happens spontaneously in response to stated objectives and in adaptation to unfolding circumstances.[102, 103]

In short: Adaptability is a function of the ease and frequency of self-organisation. For each organisational unit (at whatever level), this develops with habit-forming practice. Leaders meanwhile may need to work at it:

- Maintaining the self-discipline to allow it to happen
- Creating the opportunities for it to happen more often
- Normalising it, learning to describe outcomes without constraining solutions unnecessarily, and carefully calibrating the level of support required
- Embedding it, encourage expressions of intent in others – perhaps the most direct way to help others be successful, find meaning in their work, and exercise their own authority

None of this is peculiar to the armed forces and I am in no way suggesting that we should embrace militaristic metaphors (even less so than the factory metaphors of chapter 1 to which I have long been averse). We pay attention to these and other examples from past and recent military history for the simple reason that nowhere else have questions of culture, leadership, and organisational design been asked so deliberately over such an extended period.

In the case of leader-leader, the model doesn't even require that relationships are based on seniority, making it very easy to translate into less hierarchical settings. All it requires is some deliberately 'intentful' communication. For example, the question "*What do you intend to do?*" works in many coaching settings – it's a classic way to move a coaching conversation towards a conclusion. More specifically to a digital context, intentful communication looks like this:

- Product strategy expressed in the language of needs and outcomes – primarily the needs and outcomes of customers – the detail deferred until it is ready to be unfolded just in time
- Technology strategy expressed similarly, describing the expected qualities of solutions without constraining them prematurely
- Corporate strategy expressed in goals and objectives to which diverse parts of the organisation can contribute, perhaps innovatively

Most of the time, we're not talking big lumps of strategy, but consistent and habitual expressions of intent, sufficient to ensure that:

1. Colleagues can support, guide, check, respond to, and creatively enhance each other's work, and
2. Teams can work out how best to achieve their goals together

Needs and outcomes as catalysts for collaboration and self-organisation, in other words.

Engaged governance

It's time to keep a promise made in the previous chapter and answer a question posed earlier in this one: Who participates in the key decision-making forums?

Let's take the real example of the outside-in Service Delivery Review of the previous chapter. As you will recall, our OI-SDR meeting had five agenda items, namely Customer, Organisation, Product, Platform, and Team. Furthermore, as the meeting's facilitator, I sought to have each one led by the person or people best equipped to represent it. Here are the roles of the actual people involved in our first meeting:

- **Customer**: Product Owner and User Researcher

- **Organisation**: Service Manager
- **Product**: Product Owner (again)
- **Platform**: Tech Lead and Tech Support Lead
- **Team**: Delivery Manager (me)

Six people in total, one of them – the Product Owner – appearing twice. I should mention also that the User Researcher reported to the Product Owner, the Tech Support Lead reported to the Tech Lead, and the Service Manager (the hybrid business and technology role seen in many UK government digital services) was the most senior person present.

From this and similar positive experiences, I have arrived at my "Rule of Three", a rule of thumb that helps meetings work not just with the organisation structure but on it:

> *Design your strategy and governance meetings so that they invite the active participation of at least three levels of seniority. Include representatives from a range of different disciplines who have skin in the game and are respected for their direct knowledge of the situation.*

Not only are formal reporting lines deliberately bypassed and hierarchies thereby flattened, but there are witnesses! Information flows faster, and the quality of decisions made with it improves. No-one is there only to find fault; conversations are authentic, participants able to remain true both to the facts and to themselves.

Looking again at this OI-SDR meeting through the lens of the Spotify model, you might say that we have multiple Chapters represented, with Chapter Leads typically involved, and a sprinkling of other Chapter members too. You could imagine that the Chapter representatives also represented their respective Guilds (and often they did, though we called them *communities of practice*, and these communities extended beyond our government agency to other government agencies and departments[104]).

That's interesting, but there's an older though rather less well-known organisational governance model that has helped me formalise this better: Sociocracy, a 19th century idea tested and further developed in community and corporate settings from the mid 20th century onwards.[105]

Sociocracy operates on four principles:

1. **Informed consent**: Decisions are made neither by imposition, nor by majority vote, nor consensus, but when all *"articulated and reasoned objections"* have been addressed. When all in a group are able to accept that a decision is *"good enough for now and safe enough to try"* after

objections have been discussed and the proposal perhaps adjusted, the group's decision is made. This approach invites emotional and subconscious aspects to have a voice in decision-making along with rational considerations.

2. **Organisation in circles**: Those decisions are made in *circles*, a circle being an autonomous group of people with a shared mission and responsibility for a *domain*, an area of concern of the business. Circles may spawn new circles, lower-level circles to manage a more narrowly-defined subdomain, or higher-level *policy circles*. Importantly, people may belong to more than one circle.
3. **Double linking**: Circles overlap (Figure 31), ideally by at least two people, one the operational leader of the lower-level circle chosen by the higher-level circle, the other or others being delegated by the lower-level circle to represent them in the higher-level circle.
4. **Elections by consent**: The principle of informed consent applies to the processes of people joining circles and to the spawning process; the net effect is that everyone is in their circles by consent.

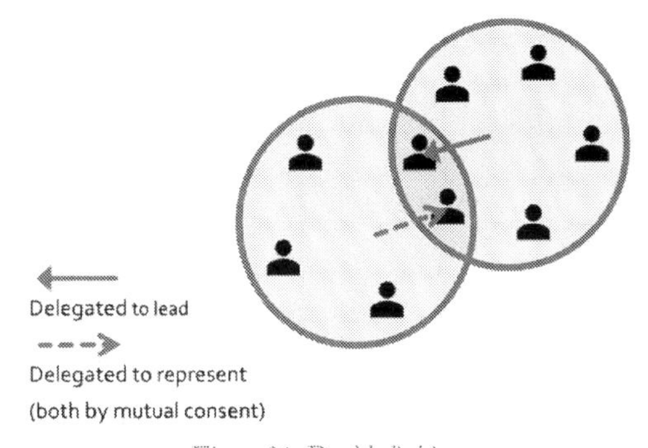

Figure 31. Double linking

In my example OI-SDR, we weren't trying to implement Sociocracy, but the correspondence is surprisingly strong. We made our decisions together, and although we hadn't formalised the decision-making process, I can't think of an occasion in which the principle of informed consent would have been violated. And recruitment worked the same way, very much a team effort. As for the circles, I can identify several of them in retrospect:

1. A Product circle, led by the Product Owner and staffed by several others, the User Researcher included
2. A Technology circle, whose members included the developers (among them a UX specialist would also have been regarded as belonging also to the Product circle), the Tech Support Lead, and

the Tech Lead

3. The Service Delivery Review group, acting as a policy circle for our combined teams and the Service as a whole, facilitated by me – the Delivery Manager – and formally led by the Service Manager
4. A Digital circle, spanning not just our digital service but several others; our Service Manager and Product Owner were both members (it should be noted that some of the other roles I mention were held on an interim basis by external contractors; these two roles were held by permanent members of staff)
5. Multiple other circles, communities of practice most notably

With the possible exception of the communities of practice, all circles double-linked to at least one other. Although we didn't know it as double-linking at the time, it seems quite reasonable to say that this was the mechanism by which my Rule of Three was met.

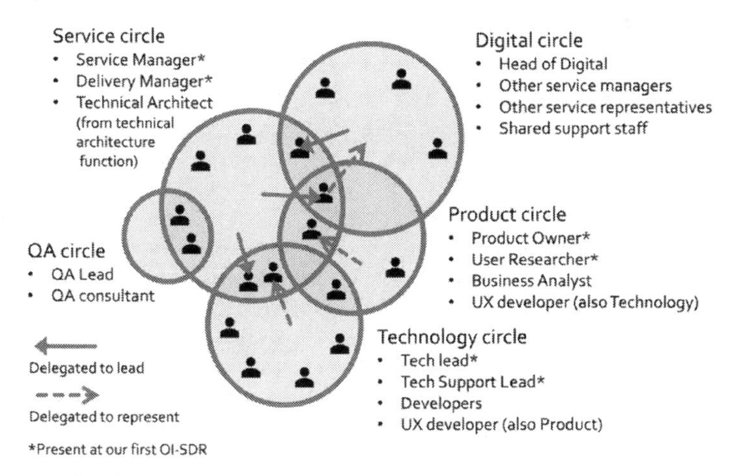

Figure 32. Our OI-SDR interpreted retrospectively as an implementation of Sociocratic governance

Not that three is an upper limit! A key part of my day when I led a global department of around a hundred people was the daily production call, held at a time convenient to Europe and the US (Asia and Australia had their own call, with a smaller number of participants dialling in from Europe). It was not uncommon for a third or more of the department to be present on the call, representing four levels of hierarchy (global, regional, team leads, and team members). Our policies would have mandated a much smaller group as a minimum, but many people joined by choice on a regular basis or occasionally through concern for a current issue.

Returning for a third and final time to examples from the military, General

Stanley McChrystal describes in *Team of Teams*[106] the daily Operations & Intelligence (O&I) meetings held during the campaign against Al Qaida in Iraq. Exercising a level of transparency that was radical for its time, these meetings involved not just multiple armed services, but also intelligence and diplomatic services from around the world. With hundreds of people in attendance by video call, he would introduce and later thank contributors by their first names, people who might be as many as eight levels his junior in rank and unused to this kind of recognition. Through this meeting, intelligence would be disseminated and cross-checked rapidly, and the operations of many autonomous teams coordinated.

Structuring participation at scale

As the scale of the organisational challenge reveals itself, the more you will find yourself dependent on the engagement of increasing numbers of people. This might sound intimidating but it needn't be, and when you get it right, the creativity unleashed can be highly energising. It can even be addictive – get it right once and you may want to do it again!

Three helpful structures that invite larger-scale participation and organisational learning are *strategy deployment* (known also by the Japanese term *hoshin kanri*), workshops of various kinds, and *Open Space* (or *Open Space Technology*, OST). We'll take these in turn.

Structure 1. Strategy deployment

Strategy deployment is a Lean approach to participatory strategy development and implementation. It can be described in quite traditional terms: top management responsible for strategy, middle management responsible for tactics, and operations staff aligning their work to those. Unfortunately, top-down descriptions such as these tend to understate both the interplay between layers and the opportunities for wide participation.

Expressing it in terms borrowed from mission command, strategy deployment could be described instead as a process of iterated *briefing* and *backbriefing*. Here, the iteration serves much the same function as it does in Agile, in that it causes an initial idea to be refined not just through discussion but also through experimentation and direct action. With this interplay sustained in multiple places and extended over time, the strategy evolves at it unfolds.

Expressed instead in Sociocracy terms it works like this:

- A challenge is formulated by a policy circle and shared for consideration by its connected circles, and so on out to operations circles

- Responses come back not just in the form of plans for action, but in new ways of thinking about the challenge, potentially causing the challenge to be reframed.

The more that circles overlap and the faster the iterations of outward and inward collaboration, the faster the strategy evolves and the closer everyone's work aligns to it. Adaptability at scale is neither top-down nor bottom-up; rather it is depends on the right choices being made at the right level at the right time, the opportunities for which are created continuously in the interplay within and between circles, based on how they operate, how they govern themselves, how they improve, and how they participate in those bigger conversations.

Structure 2: Workshops

The suggestion that a problem might be solved through a workshop isn't always welcomed, and it is certainly true that poorly-facilitated workshops can be uncomfortable and unproductive experiences for all involved. With practice however, they can be both enjoyable and highly effective.

Good workshop experiences come from good preparation. However, the facilitator's job is not to provide ready answers but to give both shape and space to the conversation. Productive group conversations depend on maintaining the right balance between constraint and freedom, a balance that varies according to the nature of the challenge, the number of people involved (anything from a handful up to hundreds), who those people are, and the power relationships that exist between them. The facilitator achieves this balance through the choice of tools employed, thereby designing and actively managing the experience. Elements of a workshop's design include the invitation, the setting of any prework, the agenda, the physical space, the way that participants are grouped, and the facilitation patterns that give the required amount of structure to each conversation or group exercise.

One rich source of facilitation patterns is *Liberating Structures*[107] – a book, a website, an app, and also a flourishing global community. I also reference Sam Kaner's classic *Facilitator's Guide to Participatory Decision-Making*[108], important for its exploration of the fundamental *diverge / converge* pattern we saw in chapter 3 in relation to Design Thinking.

With practice, some workshops (or elements thereof) become repeatable in the sense that the same structure can be taken from one context to another, deal with a different set of challenges, and quite reliably produce radically different but equally valuable outputs. For practitioners, after repeatability comes transferability, in which the ability to facilitate a repeatable workshop is transferred to other people, spreading its use across an organisation or community of practice to the benefit of many more people. A successful transfer is a demonstration of individual and interpersonal competence and leadership; for the organisation it's the opportunity for cultural, technical, and strategic ideas to spread rapidly.

After transferability comes reproducibility – a feature mentioned in chapter 4 in relation to Open Social Technologies – whereby the experience can be reproduced and perhaps improved upon outside of the control of its originator. We have seen some examples of reproducible workshops in previous chapters; another highly relevant example would be the Design Sprint[109], a week-long workshop whose goal is to identify, design, and validate a core product concept. Design sprints originated in the design community, were taken up by the startup community[110], and are now practiced across a wide variety of product organisations.

Reproducibility is a concept borrowed from science. When an experiment is reproduced, it greatly increases confidence in both the method of the experiment and the theory that the experiment was designed to test. Multiple tests under different conditions begin to establish boundaries of applicability for the method, the theory, or both.

Reproducibility can be a difficult challenge in complex environments because exactly the same experiment can be repeated under similar conditions and yet produce conflicting results. In social and life sciences, the challenge is so great that they have been described as facing a reproducibility crisis[111]. Agile doesn't face a crisis on quite that scale, but it is only now learning to effect a collective cringe when something is described inappropriately as a *best practice*, trumpeted as *'the'* Agile approach to something, or lazily given an Agile branding in the absence of a robust understanding of when, where, and how that thing should be introduced.

Structure 3: Open Space

Open Space (or more correctly, Open Space Technology) was codified in the 1980s by Harrison Owen[112]. It takes for its inspiration the hallway conversations that can often be the highlight of the conference experience; these become the metaphor for a large-scale workshop experience that is lightly but still deliberately facilitated.

Done properly, Open Space means a lot more than placing some rather lonely-looking flipcharts in a spare meeting room (the rather disappointing reality at some gatherings). It means:

1. Identifying some context – a business challenge, perhaps
2. Facilitating the emergence of a program, giving participants both the

opportunity both to propose topics relevant to the challenge and to lead their discussion in manageably-sized groups (which for practical purposes usually means sessions happening in parallel as well as in sequence over the course of the event as whole)

3. Providing the spaces in which those discussion will happen
4. Bringing together the results of those discussions – in feedback sessions or published proceedings

Some key principles guide facilitators and participants alike:

1. *Whoever comes is the right people*
2. *Whenever it starts is the right time*
3. *Wherever it starts is the right place*
4. *Whatever happens is the only thing that could have, be prepared to be surprised*
5. *When it's over, it's over (within this session)*

Last but not least, Open Space operates under a law known historically as the *law of two feet* and now as the *law of mobility*: If at any time you feel you are neither learning nor contributing, find another conversation somewhere else.

In short, Open Space is a great example of purposeful self-organisation within defined constraints, the constraints creating and enhancing the experience.

Facilitating agility in the inverted organisation

In what is sometimes referred to as *"the Agile BOSSAnova book"*[113], Jutta Eckstein and John Buck use Open Space – itself built on a metaphor – as a way to understand Agile in its broadest possible organisational context. In a similar vein, Servant Leadership and Open Space can be combined to describe opportunities for facilitation within an inverted organisation that is also highly collaborative and adaptive:

1. **Helping others to be successful**, removing impediments to those *"right people, right time, right place"* collaborations. To maximise the opportunity for effective internal and cross-unit collaboration, leaders pay attention to certain *enabling constraints*, such as meeting designs (which remain always open to experimentation), timeframes, and organisational boundaries (respected for reasons of identity and coherence but still helpfully porous and always open to adaptation)
2. **Helping others find autonomy and meaning in their work**, focussing attention on the right challenges so that collaborations are

formed with the right kind of intent. For collaborations to be entered into not just voluntarily but with passion, those clarifying challenges (including but not limited to mission, strategy, and purpose) are jointly owned and developed with a clear line of sight to external needs.

3. **Developing Servant Leadership in others**: Continuing the conversation, ensuring that learning is captured and acted on appropriately; supporting anyone who demonstrates a willingness to take the lead on any or all the above, now or in the future.

Leaders in their specific roles – digital or otherwise – can respond to these points in different ways. There's nothing unique to digital here, but the pace and complexity of the digital space is such that there will never be a shortage of opportunities in which these important leadership behaviours can be exercised.

Starting or reinvigorating the transformation

When you take those plentiful opportunities to demonstrate new leadership behaviours and then consider the number of people who might one day demonstrate them, it's not hard to understand why the word "digital" so often followed by the word "transformation", and meant in a sense that is as much organisational as technological.

There is no reliable step-by-step recipe for something as complex, multi-dimensional, and open-ended as an organisational transformation. This does not however mean that we shouldn't try both to structure our thoughts and to define a general approach.

To provide some structure to the myriad opportunities for progress, I've gathered together the reflective questions posed at the end of each chapter in **Appendix A. The end-of-chapter questions, consolidated** and also online at agendashift.com/**surveys/right-to-left**.

Including a final set of questions for this chapter, there are 50 questions in total. It needn't take long to review them however – rather than trying to answer them, just notice your gut reaction to each one. Whenever you get the sense that a question might lead to something important for your organisation, take note!

In that vein, the online version invites you to score them according to your organisation's ability to answer them and the likely quality of its answers. After completing the six sections (one for each chapter), you'll get the chance to review the complete set and to 'star' any questions that you would like to prioritise for further attention.

Before you start, note:

1. They're asked optimistically, in the sense that some of them may make some assumptions that might not apply in your organisation, so don't worry if some questions can't be answered (yet)
2. To each question there are many right answers, not just the answer suggested by the framework, model, or practice you think might have inspired it

Remember to return to the ***Taking it forward*** section immediately below once you have identified a small selection of the most pertinent questions.

Taking it forward

For just those few questions, ideally touching on a range of pressing cultural, process, framework, and infrastructural issues, and including some that have a specifically digital angle for your organisation, ask yourself:

1. What would it be like if the thing in question was *working at its best*[114] in your organisation?
2. How is that different compared to now?
3. What obstacles stand in the way?
4. What would you like to have happen?
5. Then what happens?
6. Again (and perhaps again), then what happens?

Through this process – a simplified form of *15-minute FOTO* (chapter 5) – you will have generated some outcomes, some perhaps easy to achieve, others hard. Now imagine repeating this whole process from the questionnaire onwards but this time as a participatory process, so that that agreement on outcomes can become the basis for real change.

Whether the facilitator of this process will be you or someone else, I hope that you find the prospect exciting!

Reflections

1. In your organisation, how do leaders of all kinds:
    1. Help others to be successful?
    2. Help others find autonomy and meaning in their work?
    3. Help develop these leadership behaviours in others?

2. To what extent could your organisation be described as *"existing to support those who serve its customers"*?
3. How is intent communicated within your organisation, and in which directions?
4. How does your organisation maintain authenticity in its governance activities?
5. How does your organisation encourage participation in strategy development?

---

[96] *Servant leadership: A path to high performance*, Edward D. Hess, washingtonpost.com, April 28th 2013, www.washingtonpost.com/business/capitalbusiness/servant-leadership-a-path-to-high-performance/2013/04/26/435e58b2-a7b8-11e2-8302-3c7e0ea97057_story.html

[97] *Servant Leadership: A Journey into the Nature of Legitimate Power and Greatness*, Robert K. Greenleaf (Paulist Press, 25th Anniversary edition, 2002)

[98] *Brave New Work: Are You Ready to Reinvent Your Organization?*, Aaron Dignan (Portfolio Penguin, 2019)

[99] en.wikipedia.org/wiki/George_Holt_Henshaw#First_organization_chart

[100] Or *reverse hierarchy* as the inverted pyramid is sometimes known. See https://en.wikipedia.org/wiki/Reverse_hierarchy

[101] *The Art of Action: How Leaders Close the Gaps between Plans, Actions and Results*, Stephen Bungay (Nicholas Brealey Publishing, 2011). Bungay is both a management consultant and military historian and his work is well-known inside the Lean-Agile community.

[102] *Turn the Ship Around! A True Story of Turning Followers into Leaders* L. David Marquet (Portfolio, 2013)

[103] *Leader-leader* can also be seen as introducing a Servant Leadership dimension to what management cyberneticists call the *resource bargain*. Clearly, it is futile (or worse) for managers to specify an objective without reasonable regard to the resources needed to achieve it. It's also highly wasteful if opportunities for autonomy and growth are ignored.

[104] See Emily Webber's *Building Successful Communities of Practice: Discover How Connecting People Makes Better Organisations* (Blurb, 2018). Emily was previously Head of Agile Delivery at GDS.

[105] *We the people: Consenting to a Deeper Democracy*, John Jr. Buck & Sharon Villenes (Sociocracy.info Press, second edition, 2019)

[106] *Team of Teams: New Rules of Engagement for a Complex World*, General Stanley McChristal, with David Silverman, Tantum Collins, & Chris Fussell (Portfolio Penguin, 2015)

[107] *The Surprising Power of Liberating Structures: Simple Rules to Unleash A Culture of Innovation*, Keith McCandless & Henri Lipmanowicz (2014, Liberating Structures Press); see also the website www.liberatingstructures.com

[108] *Facilitator's Guide to Participatory Decision-Making*, Sam Kaner, (Jossey-bass Business & Management Series, 3rd edition, 2014)

[109] *Design Sprint: A Practical Guidebook for Building Great Digital Products*, Richard Banfield, C. Todd Lombardo & Trace Wax (O'Reilly Media, 2015)

[110] See *Sprint: How to solve big problems and test new ideas in just five days*, Jake Knapp with John Zeratsky & Braden Kowitz (Bantam Press, 2016); this book describes the use of Design Sprints with startups in the Google Ventures portfolio.

[111] en.wikipedia.org/wiki/Replication_crisis

[112] *Open Space Technology: A User's Guide*, Harrison Owen (Berrett-Koehler Publishers, 3rd edition, 2008)

[113] *Company-wide Agility with Beyond Budgeting, Open Space & Sociocracy: Survive & Thrive on Disruption*, Jutta Eckstein and John Buck (CreateSpace Independent Publishing Platform, 2018)

[114] The phrase *"working at its best"* is inspired by Caitlin's Walker's *From Contempt to Curiosity: Creating the Conditions for Groups to Collaborate Using Clean Language and Systemic Modelling* (Clean Publishing, 2014); see also the Agendashift *True North* agendashift.com/**true-north**

Appendix A. The end-of-chapter questions, consolidated

As mentioned in the section **Starting or reinvigorating the transformation** towards the end of chapter 6, here are all 50 of the end-of-chapter reflective questions, organised by chapter. It is also available in the form of an online survey at agendashift.com/**surveys/right-to-left**.

To repeat the guidance given in chapter 6:

- A pass through this consolidated list should not take long. Rather than trying to answer them, just notice your gut reaction to each one. Whenever you get the sense that a question might lead to something important for your organisation, take note!
- In that vein, the online version invites you to score them according to your organisation's ability to answer them and the likely quality of its answers. After completing the six sections (one for each chapter), you'll get the chance to review the complete set and to 'star' any questions that you would like to prioritise for further attention.

Chapter 1. Right to Left in the material world

Lean as *"the strategic pursuit of flow, a process of organisational learning"*:

1. How do you understand those *"key moments of value creation"* that take place "on the right", happening in and resulting from the interactions between your organisation and your customers?
2. Working from right to left, how do you understand your business's value streams – the processes that culminate in those key moments?
3. More generally, how do managers in your organisation maintain an up-to-date understanding of their value streams and appropriate awareness of what is happening in them on the ground? To what extent is this based on first-hand observation?
4. Working again from right to left, how do activities in your value

streams coordinate with their counterparts upstream so that their needs are met in good time?

5. What would *"a strategy of flow efficiency"* look like for your organisation? What would sustain it?
6. Where and how do Lean's 7 *wastes* – transportation, inventory, motion, waiting, overproduction, over-processing, and defects – impact your value streams today?
7. By what mechanisms, policies, or levers are inventory, throughput, and/or time in process controlled or influenced?
8. How is psychological safety cultivated in your teams?

Chapter 2. Right to Left in the digital space

Lean-Agile as *"the strategic pursuit of flow in complex environments, the organisation placing high value on collaboration, continuous delivery, adaptation, and learning"*:

1. How does your product development organisation promote collaboration, both internal (within and between teams) and external (with customers most especially)?
2. What would your product development organisation understand by the term *working software* (or more generally, *working product*)? How compatible is that understanding with the concept of *continuous delivery*?
3. How does your product development organisation adapt to changing knowledge – product-wise, process-wise, and organisationally? How does it stimulate and capture that knowledge?
4. How does your product development organisation maintain an appropriate balance of attention across delivery, development, and infrastructure (the last of those encompassing technology, process, and culture)?
5. Working from right to left, how do you understand your product development value stream?
6. To what extent is your product development value stream anchored on the right in validation? What forms does that validation take?
7. What "upstream" activities keep the product development process fed with high value work? How do the best ideas make it to the front of the queue?
8. How do you manage work out of (as opposed to into) your product development process?

9. How do you recognise, mitigate, and address these *"flow inefficiencies"*: blocked work; stalled work; people, teams, or systems either overburdened or starved of high value work; defects; failure demand; unrealised opportunity?

10. Working from right to left across your product development value stream as it is today, which of the above flow inefficiencies would you expect to encounter first? Were you to repeat the exercise after addressing that inefficiency, what would you expect to find?

11. How many of your recurring inefficiencies could be framed as *"failures of collaboration"*? How does that framing help?

Chapter 3. Patterns and frameworks

Frameworks – *"important and easily-recognised features of the Lean-Agile landscape"* – as *"patterns to be combined in interesting and complementary ways"*:

1. How do you catalyse self-organisation and collaboration around goals? How do you cause that process to be repeated reliably in the pursuit of longer-term objectives?

2. How do you prevent left-to-right tendencies (expectations of linear and implementation-driven processes, with commitments made prematurely) from dominating in contexts where a more right-to-left approach (outcome-oriented, iterative, and just-in-time) would be more appropriate?

3. How do you ensure that teams maintain a clear sense of purpose?

4. How do you keep teams manageably small and still with the range of skills and capabilities necessary for self-sufficiency?

5. How do team-level governance mechanisms engage with those of the wider organisation?

6. Across your product development value stream, by what explicit means do you maintain attention on flow? Is work pulled into activities that have capacity available, or pushed downstream when an activity step is completed? Through what coordination mechanisms does that happen?

7. How do people, process, and technology interact to create a high-feedback environment?

8. How do you discover, identify, explore, capture, organise, and prioritise *"authentic situations of need"* and their respective outcomes? How does this understanding unfold over time? When you're starting from scratch, how do you prime the pump?

9. How do you recognise and deal with the bottlenecks and other constraints that limit the overall effectiveness of your product development process?
10. How does your organisation pursue product/market fit? By what mechanisms does it encourage experimentation and learning?

Chapter 4. Viable scaling

Some notable scaling frameworks, left-to-right and right-to-left approaches to change, and the viable organisation:

1. How do you maintain the architectural integrity of your key systems in the presence of rapid change? Similarly, how do you maintain an engaging and appropriately consistent experience for the users of your products?
2. How do you cultivate a sense of entrepreneurialism in your teams?
3. How do you maintain a level of cultural coherence across your organisation, consistent with the autonomy of each organisational unit?
4. How do you sustain *"iterated self-organisation around goals"* beyond single teams? At a technical level, how is the work of multiple teams brought together? How is the work of multiple teams aligned to shared product and business objectives? How do they collaborate?
5. What brings structure to the work of those whose remit is to encourage change in your organisation?
6. By what means are employees encouraged to participate meaningfully in change-related work?
7. When parts of the organisation are undergoing significant change, what keeps them constructively engaged with the rest of the organisation?
8. What are your organisational structures and how do they help customers and staff to identify each other, their needs, and purposes? In what ways do your organisational boundaries encourage and hinder flow?

Chapter 5. Outside in

Outside-in Service Delivery Reviews and Strategy Reviews (OI-SDR and OI-SR respectively); other outside-in tools, namely NOBL's Organisational Charter and Wardley Mapping:

1. How do you keep your reviews of current operations grounded in their proper customer and organisational context?
2. How do you keep your organisational unit's range of capabilities and perspectives aligned? Who participates in that process? What data do they bring?
3. What keeps your organisation reminded of the need for experimentation? How does it ensure that learning is captured and shared?
4. How do you decide what gets tracked? What gets discussed first?
5. For a future timeframe of your choosing, the outside-in questions:
    1. **Customer**: *What's happening when we're reaching the right customers, meeting their strategic needs?* (And: *Who are those right customers, what are their strategic needs, and why us?*)
    2. **Organisation**: *When we're meeting those strategic needs, what kind of organisation are we?*
    3. **Product**: *Through what products and services are we meeting those strategic needs?*
    4. **Platform**: *When we're that kind of organisation, meeting those strategic needs, delivering those products and services, what are the defining/critical capabilities that make it all possible?*
    5. **Team(s)**: *When we're achieving all of the above, what kind(s) of team(s) are we?*
6. Whether between the layers of the outside-in questions or within them, how are your organisation's internal contradictions dealt with? What contradictions do you face now?
7. What is your organisation's purpose? Explore that with NOBL's questions:
    1. *What do we want to change about the world and why?*
    2. *How can we use our collective skills to make change and what will the world look like when we succeed?*
8. Which of your components or capabilities might be progressed deliberately through the product maturity life cycle in order to achieve competitive advantage? What opportunities for innovation or disruption can you identify?

Chapter 6. Upside down

Servant Leadership, the inverted organisation, and participation:

1. In your organisation, how do leaders of all kinds:
    1. Help others to be successful?
    2. Help others find autonomy and meaning in their work?
    3. Help develop these leadership behaviours in others?
2. To what extent could your organisation be described as "*existing to support those who serve its customers*"?
3. How is intent communicated within your organisation, and in which directions?
4. How does your organisation maintain authenticity in its governance activities?
5. How does your organisation encourage participation in strategy development?

After reviewing this list and identifying a few of the most pertinent questions, return to the **Taking it forward** section of chapter 6.

Appendix B. My kind of...

Not a technical glossary, but gathering together some informal definitions that are especially characteristic of this book, all of them applicable in the context of digital leadership and most of them helpful in wider contexts too:

Agile (short version):

People collaborating over the rapid evolution of working software that is already beginning to meet needs

Agile (longer version):

People bringing their various skills to bear on the rapid evolution of working software that is already beginning to meet needs, working in teams that place high value on collaboration and adaptation

Digital:

Applying the culture, processes, business models & technologies of the internet era to respond to people's raised expectations

~ Tom Loosemore, *Definition of Digital*, definitionofdigital.com

Digital leader:

Anyone who sees in digital technology the opportunity to serve their customers better (meeting their needs more effectively), recognises that this may have profound implications for how their organisation should work, and is ready to help make that happen

Engagement model:

A model – of which Agendashift is an example – for how change agents do their work. To be effective in the organisational change space, an engagement model must do three things:

1. Help to structure the work of change agents – facilitators, consultants, coaches, or employees whose remit includes the encouragement of change
2. Help the client organisation engage its staff meaningfully in change-related work, inviting high levels of participation

3. Help those parts of the client organisation that are undergoing deliberate change to engage constructively with the rest of the organisation, so that all sides will thrive

Lean (short version):

The strategic pursuit of flow, a deliberate process of organisational learning

(After Modig & Åhlström's *This is Lean* and its *"a strategy of flow efficiency"*)

Lean (longer version):

The pursuit of flow as strategic imperative, an open-ended and purpose-driven endeavour that continuously and deliberately engages people at every level of the organisation in a learning process

Lean-Agile:

The strategic pursuit of flow in complex environments, the organisation placing high value on collaboration, continuous delivery, adaptation, and learning

Outside in:

Starting from outside the organisation – emphasising customers and their needs – and working inwards through layers of organisation and capability. The Outside-In Strategy Review (OI-SR) and the Outside-In Service Delivery Review (OI-SDR) are designed to create and sustain alignment (respectively).

Right to Left:

A whole-process focus on needs and outcomes coupled with a sense of *pull* from the customer side (on the right hand side of a conventional value stream map).

From the introduction: Putting outcomes before process, ends before means, vision before detail, "why" before "what", "what" before "how", and so on. It can also mean considering outputs before inputs, but give me outcomes over outputs, every time.

Rule of Three:

"Design your strategy and governance meetings so that they invite the active participation of at least three levels of seniority. Include representatives from a range of different disciplines who have skin in the game and are respected for their direct knowledge of the situation."

Upside down:

Servant leadership in inverted hierarchies:

- The organisation existing to support those who serve its customers
- Decision authority moved to those who have the most relevant information (after Marquet).

As described in chapter 6 and after Greenleaf, the servant leader's legitimacy rests on these three activities:

1. Helping others to be successful
2. Helping others find autonomy and meaning in their work
3. Helping develop Servant Leadership in others

Wholehearted:

"I choose to be in the business of helping organisations to be more *wholehearted* – less at war with themselves, their contradictions identified and owned so that they can be resolved in some satisfactory way. By way of analogy, if we improve our delivery processes by removing impediments to flow, then we improve our organisations by removing impediments to alignment."

As acknowledged in chapter 5, *wholehearted* is inspired by a quote from Christopher Alexander's *The timeless way of building*.

Resources

As mentioned in the introduction:

- This book has an online home at agendashift.com/**right-to-left**. If you have feedback of any kind you can easily reach me via this page
- There's a #right-to-left channel in the Agendashift Slack (agendashift.com/**slack**); all readers will find a warm welcome there
- If you enjoyed the book, an appreciative tweet to @asplake (me), @Right2LeftGuide (this book), &/or hashtag #Right2LeftGuide would be wonderful, thank you!

Via agendashift.com/**right-to-left** you will find links to:

- agendashift.com/**right-to-left/recommended-reading** – organised by chapter and taken from the chapter endnotes, a selection of the books and articles I have referenced in this book
- agendashift.com/**surveys/right-to-left** – as described at the end of chapter 6, the end-of-chapter reflection questions delivered in the form of an online survey
- blog.agendashift.com/**tag/right-to-left** – at the Agendashift blog (to which I post a few times a month), posts with a right-to-left theme or connection
- agendashift.com/**resources** – resources, nearly all of them published under a Creative Commons license, including Changeban (chapters 3 and 5), the Outside-in Strategy Review (OI-SR) template (chapter 4), Celebration-5W (chapter 5), 15-minute FOTO (chapter 5), A3 template (chapter 5) and pages for my Definition of Done, Rule of Three, and True North

agendashift.com/**subscribe** – join the mailing list for news of workshops (in-person and online) and new or updated resources

Acknowledgements

A special thank you to those who have graciously allowed me to include some of their work (explicitly, by open sourcing their work, or both): Tom Loosemore (Introduction), Clarke Ching (Chapter 3), Jakob Schneider and Marc Stickdorn (Chapter 3), Niels Pflaeging (chapter 4), Dan Lockton (chapter 4), Bud Caddell (Chapter 5), and Simon Wardley (Chapter 5). A big thank you also to Cara Steele and the team at the Wickes store in Chesterfield (it was they who I visited in chapter 1).

Then the review team members, whose feedback has made this a much better book than anything I could have managed on my own: Heidi Araya, John Buck, Martin Burns, Andrea Chiou, Emma Collingridge, Sharon Dale, Jutta Eckstein, Ray Edgar, Charlie Foote, Kjell Tore Guttormsen, Mike Haber, Patrick Hoverstadt, Craig Lucia, Steven Mackenzie, Angie Main, Alex Pukinskis, Karl Scotland, and Thorbjørn Sigberg. Of those, I must single out Charlie and Steven for some of the strongest comments – not just once but on multiple drafts; I'm grateful for their honesty and I hope that the finished result is suitable reward for their considerable effort.

Outside the review teams there are others whose input, support, or encouragement I would like to recognise: Phil Bowker, John Coleman, Rolf Götz, Parag Gogate, Philippe Guenet, Mo Hagar, Johnny Hermann, Adrian Howard, Sergiy Ivashyn, Jon Jorgensen, Liz Keogh, Mike Leber, Ben Linders, Daniel Mezick, Harry Nieboer, Johan Nordin, Alex Papanastassiou, Darwin Peltan, André Ribeiro, Johann Tambayah, Christophe Thibault, Vincent van der Lubbe, and Emily Webber.

It's no secret that I couldn't have done this without the enduring support and encouragement of my wife Sharon. With two books published in quick succession it must seem that I have been writing continuously for the best part of three years; for me that is no hardship but it does represent many hours engrossed in something all-consuming. Her commitment has been considerable too and for that I am enormously grateful. Sharon, you're amazing!

Dedicated to Martin Burns

In the short period between submitting the final manuscript for publication and it going to press, we learned the news of the sudden and unexpected passing of one of our review team members, Martin Burns. This was a shock not just to myself and the review team, but to the many people in the global

Lean-Agile community who counted Martin as a friend, had heard him speak, interacted with him online, or simply read something that he had written.

He is listed above as a reviewer, but I should expand here on his specific contribution to chapter 4. This, the chapter most likely to generate controversy, was the one I most wanted to get right. Martin was one of several members of the SAFe community who were able to confirm in that context what I already knew for Scrum, namely the very real tension between the left-to-right and right-to-left perspectives articulated in that chapter and the one preceding. Moreover, he was supportive both publicly and privately (including in his client engagements) with regard to outcome-orientation generally and Agendashift specifically. More than anyone else, Martin gave me the confidence to take this chapter as far as I could.

Those who knew him would not be surprised by his vocal support for two frameworks that to some eyes might appear to be in opposition. Passionate debate was in his nature; so too was an optimistic attitude towards people – he was extraordinarily *"people positive"*, to borrow (via chapter 6) Aaron Dignan's phrase. Not only was he able to keep the strongly-argued technical merits of a framework separate from implementation strategy (a separation of concerns that too many so-called experts fail at), he would trust people to do great things when given the opportunity to find a way forward for themselves.

Announcing the dedication of this book with a blog post, I finished with these words:

People like Martin don't come along every day, and it is good therefore to say thank you when they do. Martin, this one's for you..

Further tributes to Martin are linked to from mine, which you can read at blog.agendashift.com/2019/05/27/martin-this-ones-for-you/.

About the author

Mike is known to the Agile and Lean-Agile communities as the author of *Kanban from the Inside* (Blue Hole Press, 2014) and *Agendashift* (New Generation Publishing, 2018), the creator of the Featureban and Changeban simulation games, a keynote speaker at conferences around the world, and as a consultant, coach, and trainer.

Prior to founding Agendashift he was an Executive Director and global development manager for a top-tier investment bank, CTO for a late-stage startup, and (as an associate of Valtech UK) the interim delivery manager for two UK government digital 'exemplar' projects. Before and sometimes during those, a software developer (programming remains a passion).

In 2009 Mike moved with his family from the London commuter belt to the picturesque Derbyshire village of Matlock Bath on the edge of the Peak District, the UK's first National Park. He now lives a few miles further north in the town of Chesterfield with wife Sharon and daughter Florence. Sharon and Mike have two grown-up sons, Matthew and Simon.

Index

Made in the USA
Middletown, DE
01 September 2020